The Cartermill Guides

Science and Technology in Southern Europe:
Spain, Portugal, Greece and Italy

Cartermill Guides

Science and Technology in Canada
by Paul Dufour and John de la Mothe

Science and Technology in Germany
by Wilhelm Krull and Frieder Meyer-Krahmer

Science and Technology in Japan
by John Sigurdson

Forthcoming titles:

Science and Technology in China (new edition)

Science and Technology in the UK (new edition)

Science and Technology in the USA (new edition)

*Para mi amiga Carolyn Hall,
cordialmente.*
Carlos Otero
Madrid 16. IX-97

Science and Technology in Southern Europe: Spain, Portugal, Greece and Italy

Carlos Otero Hidalgo

Series Editors:
Paul R. Dufour
John de la Mothe

**Science and Technology in Southern Europe:
Spain, Portugal, Greece and Italy**

Published by Cartermill Publishing
Maple House, 149 Tottenham Court Road, London W1P 9ll, UK
Telephone: 0171-896 2424
Fax: 0171-896 2449

First published 1997

A catalogue record for this book is available from the British Library

ISBN 1-86067-185-3

© Cartermill International Limited 1997
All rights reserved. No part of this publication may be reproduced, stored in a retrieval system, or transmitted in any form or by any means, electronic, mechanical, photocopying, recording or otherwise, without the prior written permission of the publishers or a licence permitting restricted copying issued by the Copyright Licensing Agency Ltd, 90 Tottenham Court Road, London W1P 9HE.

The Cartermill Guides

Science, technology and innovation are the new currencies of global economic and social development. The rapid changes that are brought about by advances in research, new information technologies and institutional designs have had a profound impact on how we compete, do business and view the world. Corporations invest significant sums of money in developing innovative products and products to maintain competitive edge. Countries and regions around the world invest heavily in knowledge-intensive activities - from backing scientific research in universitites and colleges and supporting risk capital in high technology, to promoting a public understanding of science and fostering international cooperation in big science projects.

In this new world, timely information on science and technology is critical for effective decision-making and policy analysis. It is essential for coroprate leaders looking to gain competitive insight into innovative activity anywhere in the world. It is an asset for students looking to arm themsleves with information about current events affecting science and technology and innovation activities.

The new *Cartermill Guides to Science and Technology* are designed to address these needs. They provide decision-makers, corporate strategists, policy analysts, venture capitalists, scholars and students with detailed assessments of current research structures in strategic countries and regions around the globe. Written by leading scholars, these Guides provide detailed and up-to-date information on the main institutions and organisations (corporate, government and academic), current research strengths, funding sources, international partnerships and future trends.

The volumes in this series cover all the major performers in science and technology around the globe and provide readers with a guide to the strengths, opportunities and current and future policy directions of S&T investments and industrial innovation in selected countries and regions around the world.

Paul R. Dufour and John de la Mothe
Series Editors

CONTENTS

Preface and Acknowledgements xv
Introduction xvii

Chapter 1: Spain

Introduction 23
Legal and Institutional Framework of S&T Policy in Spain 24
    Legal Framework 24
    Institutional Structure 25
Public Research Centres 28
    Universities 29
    Public Research Bodies 31
The National Plan for R&D 34
    Mechanisms for the Development of the National Plan 35
    Objectives of the National Plan for R&D 38
National Programmes 45
    Coordination of National Programmes 51
    Coordination of Sectorial Programmes 51
    Coordination of Regional Government Programmes 53
    Other National Coordination Actions 54
European Funding for R&D 55
    European Regional Development Fund 55
International Framework 58
    European Union Framework 58
    Participation in Other International R&D Programmes 69
Technological Innovation and Development in Spanish Enterprise 72
Indicators of the Spanish Science and Technology System 74
    R&D Expenditure and Financing: General Indicators 75
    Production 80
    Scientific Environment 87
    Indicators of the Financial Environment 89
    Regional Indicators 90
Perspectives of the Spanish Science and Technology System 93
    Ideal Orientation of the Spanish S&T System. 94
    Promotion of Scientific Research 95
    Planning of Scientific Research 96
    R&D Coordination 97
    International Coordination 102
    Regional Governments 102
    Enterprise 103
    Financial Background 105

Chapter 2: Portugal

Introduction 111
Legal and Institutional Framework of S&T Policy in Portugal 112
    Legal Framework 112
    Institutional Structure 113

Public Research Centres 120
    The Role of the Public Sector 120
    Universities 121
    Description of Main Public Research Centres 123

Planning of Scientific Research: Main Public Programmes 127
    Programme for Educational Development in Portugal 127
    Specific Programme for the Development of
    Portuguese Agriculture 129
    Specific Programme for the Development of
    Portuguese Industry 129
    The *CIENCIA* Programme 131
    STRIDE 133
    PRAXIS XXI. 134

Technological Development and Innovation
in the Portuguese Businees Sector 136

Portuguese Participation in International Programmes 143
    The NATO Science and Technology Programme 143
    European Union Framework Programmes 145
    Other International Programmes 148

S&T Indicators in Portugal 150

Chapter 3: Greece

Introduction 161

Structure of the Science and Technology System in Greece 162

Insititutional Framework of the S&T System in Greece 163
    Ministry of Development 163
    National Advisory Research Council 164
    General Secretariat for Research and Technology 164

The Public Science and Technology System:
Public Research Centres and Main R&D Programmes. 166
    Public Research Centres 166
    Main Public R&D programmes: The EPET II Programme 172
    Other National Programmes 183
    Evaluation of the Results of Public Programmes 186

Public Expenditure on Scientific and Technological Research 188

Fourth R&D Framework Programme 195

R&D in the Private Sector 196

Main R&D Indicators 201

Chapter 4: Italy

Introduction 211

Institutional Framework of S&T Policy in Italy 212
    Ministry for the University and S&T Research 212
    National Council for Science and Technology 213
    National University Council 214
    National Geophysics Council 214
    National Council for Astronomic Research 214

Public Research Centres 215
    National Research Council 215
    Italian National Entity for New Technology,
    Energy and Environment 220
    Italian Space Agency 222
    National Institute for Nuclear Physics 223
    National Statistic Institute 224
    National Health Institute 225
    Other Public Research Centres 226

R&D Promoted by Ministries 227
    Ministry of Agriculture 228
    Ministry of Culture 228
    Ministry of Industry 229
    Ministry of Health 229
    Ministry for Special Measures in the Mezzogiorno 229
    Ministry of Defense 230
    Ministry of Transport 231
    Ministry of Mail and Telecommunications Services 232
    Activities of Local Administration 232
    The Operation of Public Research Centres 233

Public Research: Financing and Main Programmes 233
    National Research Programmes 235
    CNR Determinant Projects 236

Technological Innovation in Industry 252
    R&D Expenditure in Italian Enterprise:
    Fundamental Indicators 253

International Programmes: Multilateral Cooperation 263
    Main Multilateral Projects 263
    EU Framework Programmes 264

Main R&D Indicators 267

Conclusion 275

Appendices

Appendix I: General Information 281

Appendix II: Relevant Addresses 297

Appendix III: Acronyms 305

PREFACE AND ACKNOWLEDGEMENTS

In the last decade, the countries of southern Europe have profited from their highly active national R&D policies in addition to the R&D Framework and other programmes of the European Union. This has enabled rapid growth in R&D related indicators and an important evolution of the science and technology systems in these countries.

- What is the legal framework of the institutions responsible for laying down scientific policy in a country?
- Which public institutions plan and coordinate national research plans?
- What is the role of the university and private enterprise in research, innovation and technological development?
- Which research programmes are promoted by public institutions?
- Which are the main public research centres?
- What is the weight of a nation's research in the international context and how does it relate to that of other countries?
- How are the results of scientific production disseminated and what are the communication links among the different agents?

Science and Technology in Southern Europe tries to answer this series of questions. In doing so, the growing importance at international level, and especially within the European Union, of S&T systems in the south of Europe can be easily deduced.

The order of the chapters (Spain, Portugal, Greece and Italy) has its reason, given that down the years the S&T systems in the European Union Member States have adapted under the influence of the various EU bodies with identical objectives and motivation. The result is that a high number of the characteristics of these systems are common. For example, the research areas being promoted in the various countries are alike given that as the social, economic and political backgrounds are similar so are the needs that science and technology must satisfy. A double similarity therefore exists among these countries; on the one hand, institutional (homogenised institutions with similar tasks) and on the other, with respect to actions (programming). For this reason, (and to also avoid tiring out the reader) it appeared more appropriate to explore certain of these common aspects in one country and not in the others. The first chapter, Spain, focuses on the description of this type of system and the relationships among the agents. In the chapter on Portugal, the weight of private enterprise in such matters is examined and the growth in scientific research in this sector under the new and recently created legal and institutional framework is shown. The importance of the public sector in research is dealt with in detail in the chapter dedicated to Greece, given that in this country research is basically the responsibility of state institutions. Finally public research plans and their content are dealt with in the chapter on Italy.

A brief comment should be made with respect to the large quantity of statistical data. Some of this information may appear somewhat out of date, but it must be kept in mind that compilation of such information by public institutions is generally a slow process and it is difficult to find figures which are not at least two or three years old.

Monetary quantities are usually expressed in the currency of the country being dealt with. The conversion of the various currencies to a common one (US dollars, German marks or ECUs) may appear to be more convenient when comparing quantities, but this in fact is not so. Such comparison would firstly require recurrence to the purchasing power parity of these currencies in these countries and posterior conversion to a common currency. Given that this is not the objective of the book, it was therefore decided to leave figures in their original national currencies and indicate the value of the dollar and the ECU with respect to the currency in 1994.

Two annex have been included at the end of the book, considered to be of interest to the reader. The first gives a brief description of the economic background of the four countries dealt with to give the non-European reader a clear idea of their position in the world context. The second annex is a list of addresses of the main research institutions and centres which appear in the main text. Such a list was included to provide the reader with references in order to amplify his information or even initiate some type of collaboraton which could result to be of mutual interest.

Lastly, my gratitude to those who helped me in the elaboration of this book. Firstly, Dr. Tom Winston, of SPRU who encouraged me to take on the task and made possible the initial contact with the publisher. The editors, Dr. Paul Doufour and Prof. John de la Mothe for their suggestions on the structure and content of the book, and finally to those who gave their comments on the completed studies: Prof. Fernando Aldana in Spain, Prof. Augusto Medina, Prof. Antonio Fernandes and Prof. Júlio Novais in Portugal, Prof. Antonia Moropoulou in Greece and Dr. Angelo Guerrini in Italy.

Special mention should be made of Javier Macías, consultant with Estudios Institucionales, for his help in the compilation and revision of all the data which appears in this book and to Edel McLaughlin, also of Estudios Institucionales, for her cooperation in the translation.

INTRODUCTION

Underlying the concept of the European Union is the idea of creating an economy and society in the old continent capable of confronting the commercial challenges outlined by the other world powers. One of the factors, although not the only one, which permits such a challenge to be succesfully confronted is an appropiate level of development in both research and technological progress.

We live in a world in which the new technologies, and in particular those of information and telecommunications, are having a profound impact on the economies and structures of societies. The globalisation of the financial markets and trade in general, and the increasing restructuring of the production systems is in part a reflection of the impact of these technologies. Within this framework, international competitiveness, and in consequence employment, depends less and less on traditional factors such as availability of primary resources or cheap labour; the new generating areas of resources are based on knowledge, and their strategies on the continuous creation of technological innovation.

What is the balance of the European Union with respect to its main competitors? Three key aspects should be pointed out. The first deficiency observed is that the level of resources used is comparatively smaller. The USA invests 2.8% and Japan 3% of its GDP compared to 2% of the average GDP invested by the EU. The second aspect refers to the lack of coordination at various levels in European activities related to research and technological developement. This insufficient coordination occurs mainly in national research policies. Lastly, and perhaps the most serious deficiency, is that the European research system, in comparison with its competitors, is not efficient when transforming technological advances into industrial results and commercial success.

The solution to these problems is definitively within our power and is basically to be found in the reanimation of growth in this so vital sector of the common economy. A change of attitude or mentality in the scientific and research society as a whole is also called for to ensure that the results of its activity are converted into new technologies, satisfying the needs both of the new markets being opened up and those of a society which demands continuous improvements in its well-being and standard of living. In general our weakness is not our scientific production, which is of high quality, but the use that we make of it.

Such statements can not be made with out considering the European Union as a single entity; in spite of the fact that it is made up of individual countries with different characteristics and levels of developement, it has a common social and cultural origin. The question should therefore be posed regarding the weight of each country with respect to scientific research in the EU as a whole. The answer which would quickly come to the mind of many would be that the contribution of northern and central Europe is substantially greater than that of the south. However, this situation which is a clear reality, is beginning to change. For the last decade or so, Spain, Portugal, Greece and Italy have been making a special effort to converge with the EU average. This fact, which is closely tied to the development of these economies to raise them to the same level as their EU partners, is not unaware of the need for adequate development of the science and technology systems. If knowledge is to become a fundamental weapon in competitivenes, the national research systems must promote national scientific production of quality and not take refuge in technology transfer from outside, unless as a driving-force for future domestic development.

A great part of the burden of science and technology systems in the countries of southern Europe is borne by private enterprise. Surprisingly, it is in this aspect where the work of the different governments acquires greater importance, not only in the promotion of research by means of the different initiatives but also to channel effort and coordinate activities, in such a way that the dissemination of scientific results quickly and efficiently reach the possible user. In southern Europe, these public systems are already consolidated and have for some time been endowed with the appropriate legal frameworks, while their public research institutions and organisations have been satisfactorily carrying out their functions. Perhaps the most important fact is that the channels of communication between public research, the university and enterprise are already established and are now beginning to show the first, and expected, results.

Much remains to be done, especially for the countries of southern Europe. The foundations of the science and technology systems have already been laid but must now be developed in order to achieve the goals expected of them. This will happen when these societies become aware of the importance of scientific and technological development and its immense potential, a potential which is no longer only industrial and commercial but also the potential of its social impact at all levels, especially on employment.

SPAIN

Chapter 1

SPAIN

INTRODUCTION

In recent years, Spain has undergone a rapid process of incorporation to the international world, fulfilling the role which corresponds to its historical tradition, cultural image and level of development.[1]

This statement was made by the President of the Spanish Government in 1988 in reference to the S&T system and indicates the reality of scientific research in Spain at present. Until recent times, such a statement would have been considered utopian. Throughout history, research and technological development in Spain has been subject to a series of deficiencies, largely due to lack of stimuli and appropriate instruments for development. The country's contribution to scientific progress in general has therefore been quite scarce, except for the appearance of certain sporadic figures, especially at the beginning of this century.

The present situation has changed, and the change is due to Spain's joining the European Union. Since 1986, the nation as a whole has had to come to terms with a system of competition and the gradual elimination of traditional, governmental protectionism. In the light of this new situation, public administration, through its institutions, enterprise and society in general, became aware of the need to develop a new S&T system in line with these new circumstances, in such a way that the country was endowed with the level of competitiveness necessary to enable it to converge with its new EU partners.

As will be seen throughout the course of this book, the three fundamental aspects on which institutions and organisations responsible for the development of the Spanish S&T focus are the following:

- Development of Scientific Research
- Planning of Scientific Research
- Coordination of Scientific Research

The Law for the General Coordination and Development of Scientific and Technical Research aims to correct traditional shortcomings and constitutes the legal framework for the development of the system.

The provisions of this Law have resulted in the State establishing a series of institutions where promotion, planning and coordination actions have been centralised through the creation of the National Plan for R&D. These institutions, in addition to the National Plan, are now consolidated and respond to the expectations created in the scientific environment despite some deficiencies.

Against such a background, reference can now be made to an authentic S&T system, institutionally well organised and with certain expertise as regards actions. The next step is probably greater awareness of Spanish society regarding the importance of technological development and a gradual increase in investment in such areas. Expectations of the system converging with the EU average will therefore become reality, and will hopefully take place in the first decade of the 21st century.

LEGAL AND INSTITUTIONAL FRAMEWORK OF SCIENCE AND TECHNOLOGY POLICY IN SPAIN

Legal Framework

Law 13/1986 of the General Promotion and Coordination of Scientific and Technical Research, better known as the Science Law, makes up the legal framework of the S&T system in Spain, and lays out the principles for reform of the system and the bodies responsible for carrying it out. New mechanisms for action are determined, consisting of different organisations, while participation and coordination among them are guaranteed.

Examination of the motivation behind this Law, establishes that it must correct the "traditional wrongs" of scientific production and technique in Spain: these being insufficient endowment of resources, disordered coordination and administration of research programmes, and limited Spanish participation in the technology processes in which other industrialised countries are involved.

Figure 1-1: Main Public Insitutions of the S&T System

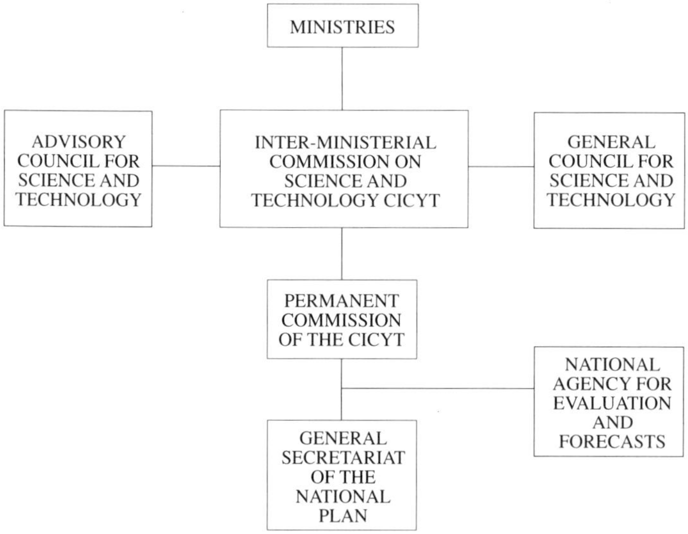

```
                        ┌─────────────┐
                        │ MINISTRIES  │
                        └──────┬──────┘
                               │
  ┌──────────┐          ┌──────┴───────┐          ┌──────────┐
  │ ADVISORY │          │INTER-MINIST. │          │ GENERAL  │
  │ COUNCIL  ├──────────┤ COMMISSION ON├──────────┤ COUNCIL  │
  │   FOR    │          │  SCIENCE AND │          │   FOR    │
  │ SCIENCE  │          │  TECHNOLOGY  │          │ SCIENCE  │
  │   AND    │          │    CICYT     │          │   AND    │
  │TECHNOLOGY│          └──────┬───────┘          │TECHNOLOGY│
  └──────────┘                 │                  └──────────┘
                        ┌──────┴───────┐
                        │  PERMANENT   │
                        │  COMMISSION  │
                        │ OF THE CICYT │
                        └──────┬───────┘
                               │                  ┌──────────┐
                               │                  │ NATIONAL │
                               │                  │AGENCY FOR│
                               ├──────────────────┤EVALUATION│
                               │                  │   AND    │
                        ┌──────┴───────┐          │FORECASTS │
                        │   GENERAL    │          └──────────┘
                        │ SECRETARIAT  │
                        │   OF THE     │
                        │   NATIONAL   │
                        │    PLAN      │
                        └──────────────┘
```

Source: SGPN (General Secretariat of the National Plan)

The Law proposed the creation of the National Plan for R&D (hereforth referred to simply as the National Plan) whose mission is basically that of promoting, planning, coordinating and funding scientific research projects. Since 1988, together with the bodies in charge of its development and execution, this Plan has oriented Spanish S&T toward strategic objectives.

The Science Law places special emphasis on the development of research links among public research centres, universities and enterprise (Article 5) and on the fundamental role of regional government in propelling regional R&D actions (Article 12).

The Laws of University Reform (1983), Patents (1986) and Intellectual Property (1986) are complementary legislation.

Institutional Structure

As already noted, the Science Law establishes the necessary institutional mechanisms for the development of the S&T system in Spain. Such institutions are shown in Figure 1-1.

Interministerial Commission on Science and Technology

The Inter-Ministerial Commission on Science and Technology (*Comisión Interministerial de Ciencia y Tecnología - CICYT*) is the entity with responsibility for the preparation, planning, coordination and follow up of the National Plan. The *CICYT* is presided over by the Minister for Education and Science. All Ministries in some way related to research and development have representation (Economy and Treasury; Defence; Agriculture and Fisheries; Health; Education and Science; Culture; Industry; Presidency of the Government; Trade and Tourism; and Public Works). The Commission's main objective is to define the science policy to be implemented in the country and to establish the most appropriate mechanisms for its development. It coordinates Spanish participation in international, bilateral or multilateral projects, both at EU level and beyond, incorporating these projects into the structure of the National Plan and ensuring adequate scientific collaboration with the Centre for Industrial Technological Development (*Centro para el Desarrollo Tecnológico Industrial - CDTI*).

The *CICYT* is responsible for assigning economic resources to the National Fund for the Development of Scientific Research, the main instrument of public R&D financing. It also negotiates agreed private funds, dedicated to the different programmes making up the National Plan.

Other tasks are the orientation of training policies for researchers at all levels, the proposal of measures for the development of employment as well as facilitating mobility in the areas of research and production.

Lastly, this Commission presents an annual report to the Government and Parliament, containing an evaluation of the execution of the National Plan as well as recommending proposals and modifications to be introduced.

General Council for Science and Technology

The President of the General Council for Science and Technology is also that of the *CICYT*. The Council's mission is the promotion and general coordination of R&D activities among the regional governments, and between these and central administration. The Council consists of representatives from each regional government and *CICYT* government appointed members.

Apart from this first objective, another task to be highlighted is that of informing the National Plan to ensure correct use of available resources, propose research programmes and projects to the regional administrations, promote the exchange of information between these and central administration, and to administrate documentation on the research plans promoted by public bodies.

Advisory Council for Science and Technology

The Advisory Council for Science and Technology is responsible for promoting the participation of the scientific community and economic and social agents in the preparation, follow-up and evaluation of the National Plan. It establishes links between programmed scientific activity and social needs and interests. One of its functions is the preparation of proposals and laying out of objectives to be included in the National Plan, as well as reporting on its economic and social repercussion. At present it is presided over by the Minister of Industry, and is made up of scientists, members of business and scientific associations, trade union representatives, as well as representatives of the Ministry of Industry and Energy (*Ministerio de Industria y Energía - MINER*) and the *CICYT*.

General Secretariat of the National Plan

The General Secretariat of the National Plan (*Secretaría General del Plan Nacional - SGPN*) comes under the control of the Permanent Commission. Its main function is to provide support to the Inter-Ministerial Commission on Science and Technology. It is in charge of the coordination of the technical, economic and administrative management of the National Plan, as well as its activities, and provides the information necessary for the carrying out and fulfilment of the Plan.

CICYT Permanent Commission

The Permanent Commission of the *CICYT* is in charge of drawing up the National Plan, in its scientific-technical context, in addition to supervising its evaluation and follow up. Members are government appointed members of the *CICYT*, although the Commission may also temporarily appoint scientific personnel from ministry departments, regional governments, universities, public research bodies, etc., deemed necessary to carry out its activities. At the moment, the means available to it are established by the Ministry of Education and Science. It is presided over by the Secretary of State for Universities.

National Agency for Evaluation and Forecasts

Like the *SGNP*, the National Agency for Evaluation and Forecasts (*Agencia Nacional de Evaluación y Prospectiva - ANEP*) comes under the control of the Permanent Commission. It studies and evaluates, from a technical viewpoint, projects presented by public research centres or enterprise. Evaluation is focused on technical aspects of the project, that is to say its scientific quality, and their conformity to the guidelines laid down by the National Plan. The *ANEP* also carries out studies in themes related to science and technology in areas of responsibility of the *CICYT*. The evaluation method and type of analysis used by its 700 specialists, is that of peer review.

Parliament - Senate Commission

The main mission of this Commission, formed by 22 parliamentarians and 16 senators, is the follow-up and political evaluation of the activities and results of the National Plan. The results are published in an annual report. Responsibility is also assumed for the political analysis of the different policies applied in any area of science and technology.

In addition to the entities outlined above, there exist others with more specific assignments within the area of promotion, planning and coordination of the Spanish S&T system. The management of certain aspects of the National Plan has been transferred to such organisations. The most important are outlined below.

Centre for Technological Industrial Development

The Centre for Technological Industrial Development (*Centro de Desarrollo Tecnologico Industrial - CDTI*) comes under the jurisdiction of the Ministry of Industry and Energy. By decision of the *CICYT*, the *CDTI* is responsible for the co-administration of programmes whose objectives are primarily industrial. It also manages R&D cooperation between enterprise and public research centres.

The objectives of the *CDTI* as co-adminstrator are as follows:

- To ensure adequate returns as regards technology transfer and information, and financing.
- To strengthen the participation of Spanish companies in the European Single Market.
- To defend the national interests reflected in its work plan.

These objectives are achieved due to the fact that the *CDTI* is involved in the: preparation of programmes, providing aid to companies in all the project phases;

programme administration; coordination of actions and interests; and dissemination of results.

This role of the *CDTI* as co-administrator also extends to funds provided by the EU through the R&D Framework Programmes related to industry.

General Directorate of Scientific and Technical Research

The General Directorate of Scientific and Technical Research (*Dirección General de Investigación Científica y Técnica - DGICYT*), dependent both administratively and financially on the Ministry of Education and Science, administers: the Sectorial Programme of Teacher Training and Improvement of Research Personnel; the Sectorial Programme of General Promotion of Knowledge; and the National Programme of Training of Research Staff.

PUBLIC RESEARCH CENTRES

Before going into a description of the main public research centres, it is necessary to outline the relationship that exists among the group of institutions, centres and national programmes making up the main body of the Spanish S&T system. This can be seen in the flowchart where such a relationship as well as the main channels of coordination are indicated (marked with heavy line).

Figure 1-2: The Spanish S&T System: Links

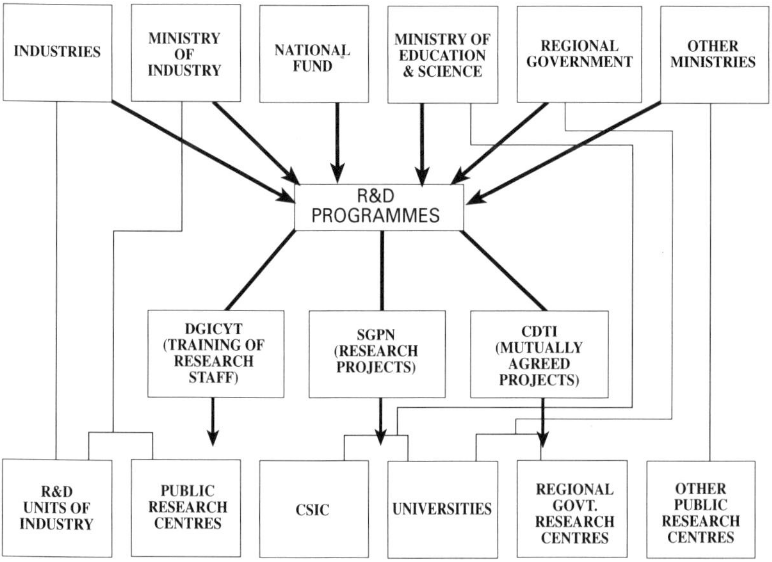

Source: SGPN

Universities

The greater part of research personnel in Spain is concentrated within the universities. According to the National Statistics Institute (*Instituto Nacional de Estadística - INE*), more than 50% of Spanish scientists carry out their work within the university framework. In addition to their education linked tasks, university centres develop a very important percentage of national scientific projects. In recent years, support to different socio-economic sectors through research has grown substantially in such centres.

Further statistics provided by the *INE* show that, in 1992, R&D expenditure in universities was more than 150 billion pesetas. This figure makes up more than 28% of total national investment in R&D. In recent years, this figure has increased, reaching a level of around 30%. In the following figure, the budget breakdown by the activities developed in these centres can be seen.

Figure 1-3: Research expenditure in universities according to area *

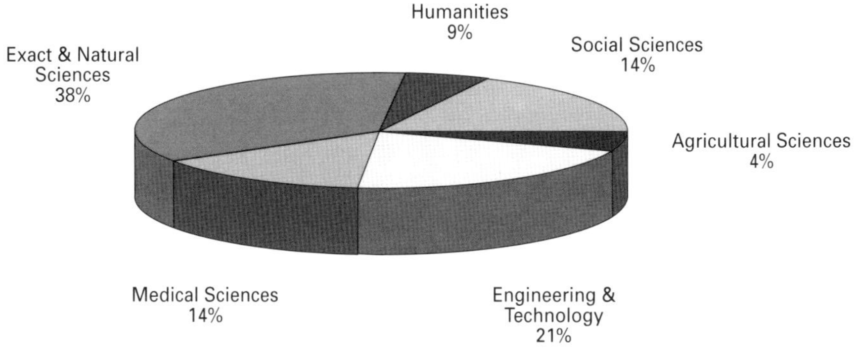

- Humanities 9%
- Social Sciences 14%
- Agricultural Sciences 4%
- Engineering & Technology 21%
- Medical Sciences 14%
- Exact & Natural Sciences 38%

Total: 32,893 Million pesetas

Source: INE, 1994

Research sectors related to social sciences and exact and natural sciences make up more than 60% of the budget, while the remaining sectors, perhaps of a more practical nature if only from an economic viewpoint, make up only 40%. This data leads to the conclusion that university research is possibly not adequately oriented. Assuming that Spain is in a process of industrial modernisation, it is certainly contradictory that only 17% of the total budget is dedicated to industrial and technological matters.

As regards research personnel, the tendency observed corresponds with that of funds distributed according to activity. In 1994, university based researchers made up 54% of the overall national figure. Breakdown by scientific area is shown in Figure 1-4.

There are 45 public universities in Spain, including the National University for Distance Learning and the *Universidad Internacional Menéndez Pelayo*. In

---

* Average exchange rates in 1994: 134 ptas=1 $; 82.6 ptas=1 Deutschmark; 158.5 ptas=1 ECU.

addition, there are 7 private universities, a number increasing considerably due to the greater liberalisation of higher education in the private sector.

Figure 1-4: Distribution of Researchers in Universities by Scientific Area

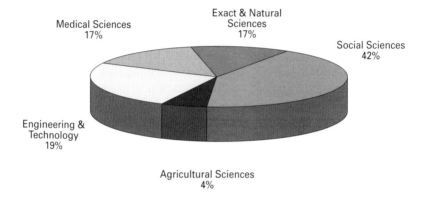

Total: 28,591 researchers

Data: INE, 1994

Figure 1-5: Organisation Structure of the Main Public Research Centres

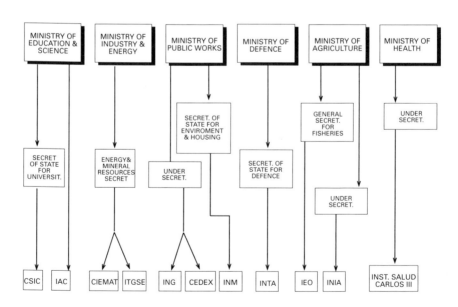

Source: SGPN

Public Research Bodies

Public research organisations are associated to a specific ministerial department, and are responsible for the development of the sectorial programmes corresponding to their ministry. They also participate in National Plan projects through the corresponding annual calls for application. At present, eleven such bodies exist: two in the Ministry of Education and Science (*CSIC* and *IAC*), two in the Ministry of Industry and Energy (*CIEMAT* and *ITGE*), three in the Ministry of Public Works (*IGN, CEDEX* and *INM*), one in the Ministry of Defence (*INTA*), two in the Ministry of Agriculture and Fisheries (*IEO* and *INIA*) and one in the Ministry of Health and Consumption (*Instituto de Salud Carlos III*).

Links between these organisations and the corresponding ministerial departments are shown in Figure 1-5.

Ministry of Education and Science

Higher Council for Scientific Research

The Higher Council for Scientific Research (*Consejo Superior de Investigaciones Científicas - CSIC*) is the largest public research centre in Spain. The projects which it carries out, cover almost all fields of science, especially those related to humanities: biology, chemistry, physics, agrarian sciences and mathematics. At present it employs more than 1,800 researchers working in 86 research centres. In 1994, its budget was 31,567 million pesetas, and resources valued at 20,039 million pesetas were generated, supposing a degree of self-financing of 39%. The *CSIC* has a series of support departments (calculus, documentation, information) and central administration services. Cooperation agreements are negotiated with universities, regional governments and foreign research centres.

Astrophysics Institute of the Canary Islands

Scientific research projects in astronomy and related fields are developed in the Astrophysics Institute of the Canary Islands (*Instituto Astrofísico de Canarias - IAC*) which depends on the Regional Government of The Canaries, state administration, the *CSIC* and the *Universidad de la Laguna*. Due to the privileged position of the Canary Islands, the *IAC* has been able to bring together a large number of foreign researchers and it has promoted cooperation agreements with various international research bodies in the area of astrophysics. Main activities are carried out in the *Observatorio Roque de los Muchachos* and in the *Observatorio del Teide*, both belonging to the *IAC*. The budget in 1994 was 1,138 million pesetas. It has a research staff of 114.

Ministry of Industry and Energy

Technological Geological and Mining Institute of Spain

In the Technological Geological and Mining Institute of Spain (*Instituto Tecnológico Geominero de España - ITGE*), the majority of research projects developed are related to areas of mining: hydrogeology, geotechniques, research of marine resources, environmental geology, mining safety and geologic risk. In 1994, the budget was 3,310 million pesetas. The number of researchers is 190.

Centre of Energy, Environmental and Technological Research

Most of the scientific projects of the Centre of Energy, Environmental and Technological Research (*Centro de Investigaciones Energéticas, Medioambientales y Tecnológicas - CIEMAT*) are related to the solution of energy problems. The studies carried out are aimed at making the best possible use of available energy sources and research into new forms of energy (renewable energy, fusion, etc.). The commercial nature of this institute results in its main objective being the improvement of the competitiveness of Spanish industry. It interacts with enterprise, R&D programmes, and central and regional government. It is organised in four independent institutes: Renewable Energy, Environment, Nuclear Technology, Basic Research. *CIEMAT*'s budget in 1994 was 9,755 million pesetas, self-financing reaching 35%. A total of 490 researchers work in the Centre.

Ministry of Defence

National Institute of Aerospace Technology

Due to the nature of projects, related to the areas of aeronautical and aerospace technology, carried out in the National Institute of Aerospace Technology (*Instituto Nacional de Técnica Aeroespacial- INTA*), there is a high degree of international cooperation. Main collaborators are the EU, NASA (National Aeronautics and Space Administration) and ESA (European Space Agency). The budget in 1994 was 16,672 million pesetas. It employs 1,400 researchers.

Ministry of Agriculture, Fisheries and Food

Spanish Institute of Oceanography

The Institute's research activities are centred on areas related to oceanography, environment, fisheries, biology and marine geology, marine ecology, marine contamination, oceanographic physics and chemistry, etc. The 1995 budget was 3,288 million pesetas. 150 researchers are employed here.

National Institute of Agricultural and Food Research and Technology

The National Institute of Agricultural and Food Research and Technology (*Instituto Nacional de Investigación y Tecnología Agraria y Alimentaria - INIA*) focuses its research on topics related to agriculture, forestry development and agricultural industry. It is responsible for the development of the Sectorial Plan of Agricultural R&D. The results of studies are disseminated in such a way as to ensure that they are transformed into new advances and tendencies in these sectors, training of new researchers and the development of international cooperation agreements.

The 1994 budget was 5,032 million pesetas, of which 2,035 were dedicated to the sectorial programme. Total staff is 726, with 519 dedicated to research.

Ministry of Health and Consumption

Carlos III Health Institute

Both the Ministry of Health and Consumption and regional government health services can make use of this Centre as scientific-technical back-up. Activities developed here are related to health, clinical, bio-medical, medicinal-biological R&D as well as food and environmental security. The Institute: coordinates the activities of various national research centres; promotes R&D in health through its research units in the National Health System; evaluates health technologies through the Evaluation Agency for Health Technology (*Agencia de Evaluación de Tecnologías Sanitarias*); is responsible for training of personnel, especially in areas related to public health and administration. It shares responsibility with the National Centre of Epidemiology for epidemic surveillance. In carrying out its tasks, the Institute collaborates with a number of national centres: Microbiology, Virology and Sanitary Immunity, Pharmacology, Environmental Health, Cellular Biology and Retrovirus, and Food.

In 1995, the budget was 13 billion pesetas.

Ministry of Public Works and Transport

National Geographical Institute

The main mission is the cartography of the national territory. It elaborates R&D plans related to geodesy, cartographic training, astronomy and geophysics, cartographic processes, meteorology, seismology, etc. The main research centres depending on this Institute are: the National Seismic Network, Geomagnetic Observatories, the Astronomical Centre of Yebes and the Observatory of Calar Alto. Of the 4,526 million pesetas in the 1993 budget, 1,268 million was dedicated to research. The number of research staff is 108.

National Institute of Meteorology

Negotiation, coordination and administration of activities related to meteorological R&D in Spain is the responsibility of the National Institute of Meteorology (*Instituto Nacional de Meteorología - INM*). To carry out its tasks, the *INM* has 15 meteorological centres distributed throughout the entire national territory. Activities are carried out in relation to weather forecasting and the study of climatology. It should be pointed out that this Institute has undergone high growth in recent years, in line with the increase in funding. It has also increased its number of support centres.

In 1993, the budget was 9,379 million pesetas, of which 1,042 million were dedicated to research activities. It employs over 200 meteorologists.

Centre of Public Works, Studies and Experiments

Contrary to other public research centres, the Centre of Public Works, Studies and Experiments (*Centro de Estudios y Experimentación de Obras Públicas - CEDEX*) is autonomous and clearly commercially oriented. It is associated to the Under Secretariat of Public Works. Main research functions are in matters related to civil works (ports, coasts, oceanography, hydrology, environment in public works, communications infrastructure, geotechnique, etc.). As part of these basic functions, importance is attributed to technology transfer, high level consultancy and the training of research personnel. Of the 1994 budget (5,235 million pesetas), 2,019 were devoted entirely to tasks related to research and development. The number of researchers is 690.

THE NATIONAL PLAN FOR R&D

The first article of the so called "Science Law" established the National Plan for Scientific Research and Technological Development, the objective of which is the promotion, coordination and programming of R&D in Spain. From the moment of its creation, the National Plan became the fundamental instrument for the development of scientific policy in Spain.

The objectives of the S&T system are laid out in the National Plan, in particular, those considered to be of high priority for national socio-economic development. Priorities are established and the resources and funds necessary for their development are detailed.

These objectives are:

- Progress of knowledge and advance of innovation and technological development.
- Conservation, enrichment and good use of natural resources.
- Economic growth, creation of employment and the improvement of working conditions.

- Development and strengthening of the competitive capacity of industry, trade, agriculture and fishing.
- Development of public services and, especially those related to housing, communications and transport.
- Development of health, of social well-being and quality of life.
- Strengthening of national defence.
- Protection and conservation of the artistic and historical patrimony.
- The development of artistic creation and the advance and dissemination of culture.
- The improvement of the quality of teaching.
- The adaptation of Spanish society to the changes brought about by scientific development and new technologies.

The definition of the programmes making up the National Plan will take into account the following:

- Social and economic needs in Spain.
- Existing human resources and materials in the Spanish scientific and technological community and their future needs.
- Available economic and budgetary resources, as well as the need for regular financing in order to maintain and promote high quality scientific and technical research.
- The need to reach a high capacity in science and technology.
- The convenience of accessing external, quality technologies by means of selective incorporation processes, suited to the development of Spanish scientific and technological capacity.
- Human, social and economic repercussions brought about as a result of scientific research or its technological application.

The following sections outline the instruments used by the National Plan to develop its programmes, in addition to the public organisations where the main research activities are developed.

Mechanisms for the Development of the National Plan

As already indicated, the National Fund for Scientific and Technological Development is the budgetary instrument used to finance the various national R&D programmes, distributed according to themes and activity. This section describes the structure of the National Plan; its axis of activity or mechanisms for the development of its objectives.

A primary subdivision, according to the level of public administration responsible, would be:

- National Programmes
- Sectorial Programmes
- Programmes in the framework of regional governments.

National programmes mainly follow high-priority guidelines of national interest and are usually directed toward preferred fields. Normally, they receive financing from the National Fund and are multi-institutional. They follow through on the entire project, from the initial phases to culmination, regarding results, or products or innovative processes that can be derived.

Programmes, which due to their nature, require special treatment in their administration or theme are included within the category of National Programmes as "horizontal" programmes, which in order to be developed, must be framed in a very ample multi-disciplinary context and demand special effort in planning; in brief, areas such as communications, documentation or training.

The following type of programmes referred to are Sectorial Programmes. These are promoted by a certain organisation or ministerial department which takes charge of their administration. As a large number of such programmes are significant or require multi-disciplinary treatment, they may be included in the National Plan at the request of the promoting body. These programmes may be carried out within the entity in question or a public research centre.

Lastly, the third type of programmes are those carried out within the framework of a regional government, subsequently the promoter. At the request of regional organisations, certain R&D programmes can be included in the National Plan, such as those which are of special interest to the nation as a whole or demand coordination of effort at national level. Due to their regional-national character, financing is usually shared between the *CICYT*, through the National Fund, and the regional body involved.

A second subdivision, based on the objectives of the programme or project, can be established:

- Research Projects
- Projects for Training of Research Personnel.
- Integrated projects
- Scientific-technical Infrastructure.
- Special Actions.
- *PETRI* Actions.
- Mutually Agreed Projects.

Research Projects

Research projects are the central axis of R&D activity. The development of such projects within the National Plan covers the initial proposal right through to project end. The terms of execution are usually three years (length of each National Plan) and they are developed in public research centres or other non-profit making research bodies. Financing by the National Fund includes payment of research personnel and the acquisition of the material necessary to develop the project. Gradual modernisation of public research centres, which until recently lacked suitable equipment and instruments, has been brought about through these research programmes.

Training of Research Personnel

This type of project is dedicated to the training of new research personnel and perfecting the knowledge of existing personnel. They usually consist of courses or seminars carried out in Spain, although part of the financing is destined for the training of specialists in prestigious centres abroad or recruitment of foreign researchers to enable them to develop part of their activity in Spain. These projects also promote mobility of research between public research centres and enterprise. Like the Sectorial Programme for the Training of Teachers and Improvement of Research Personnel, these programmes have led to a substantial improvement in human resources in the Spanish S&T system.

Scientific-Technical Infrastructure

These types of programme consist of the provision of the necessary infrastructure and equipment to research projects. Financing provided by the European Structural Funds of the European Union should also be taken into account here as they have greatly contributed to the improvement of scientific-technical infrastructure in Spain. Both types of funding are usually dedicated to the acquisition of large instruments or the renovation of those that have become obsolete. Also included within the investment projects, is the equipment necessary to provide laboratories with the necessary instruments, maintenance, etc. It should also be pointed, that these funds, negotiated through the National Plan, have encouraged investment by other entities involved in the programmes. In some areas, especially those related to public works, maintenance contracts and service etc., a certain degree of co-financing by other organisations has been achieved.

Special Actions

Special actions provide support to different scientific projects and are of a reduced time span, consisting of the organisation of courses, seminars, conferences, coordination of proposals for international projects, etc. Given that it is often difficult to foresee needs and the number of such actions, they must be flexible and quick to act.

Integrated Projects

Integrated projects are projects demanding multi-disciplinary coordination due to their nature. They are projects wide ranging projects, which require the participation of different specialists and public research centres, and usually show a high degree of participation by enterprise throughout the course of their development.

PETRI Actions

The objectives of *PETRI* (*Programa de Estímulo a la Transferencia de Resultados de Investigación* - Programmes to Stimulate the Transfer of Research Results) is to promote relations between public research centres and enterprise; that is to say, ensure that the research carried out in public bodies continues to be developed up to the point where it may be applied by industry. These actions constitute an essential part of the National Plan for R&D, given that an S&T system would not make sense if a high proportion of results had no industrial application that brought about improved competitiveness.

Mutually Agreed Projects

The objective of mutually agreed projects is the development of R&D activities in enterprise. The National Fund takes charge of financing research projects in enterprise, provided that the project corresponds to the guidelines established by National Plan objectives. Normally, financing consists of interest free credits, the objective of which is to reconcile scientific interests with purely economic interests.

As pointed out beforehand, of the institutions in charge of the development and coordination of the National Plan, the *CDTI* is the entity responsible for the evaluation of the technological and economic interest of company projects. *CDTI* also determines if projects conform to the guidelines established by the National Plan. If the programme is international, both the *CDTI* and *CICYT* will try to ensure that there are adequate technological and industrial returns. Lastly, the *CDTI* also coordinates enterprise projects which require cooperation by a public research centre.

Objectives of the National Plan for R&D

Research and Development

A fundamental need of any process of scientific and technological development is adequate financial support. A high-priority objective of the National Plan is the capitalisation of the Spanish S&T system, in order to ensure that research centres, public and private, developing activities in relation to National Plan programmes, have sufficient resources to remunerate research personnel and the necessary infrastructure and equipment at their disposal to develop their activities.

The strengthening of R&D activity cannot be indifferent to detailed programming; budget entries must focus on scientific and technological areas conforming to high-priority programmes, due to economic and technological interest as well as social interest. Neither is it convenient to forget that economic support must be provided for basic research, which in spite of having less immediate application, is of great interest in the long term. Breakdown of the National Fund, according to its axes of activity and scientific and technological areas, allows the destination of public R&D funding to be determined and controlled, as well as adapted to the above mentioned interests.

It must also be kept in mind that appropriate planning through the National Plan has allowed a substantial increase in budget entries dedicated to R&D, thus allowing the capitalisation of the Spanish S&T system.

Overall 200 research projects have been subsidised through the National Plan with investment up to 1994 reaching 86 billion pesetas. 35 billion pesetas have been dedicated to the financing of mutually agreed projects, scientific collaborations between enterprise and public research centres or universities. The contribution of enterprise should be added to this figure in order to calculate the total sum dedicated to such collaborations. Other funding sources exist in this area (such as *MINER*) which provides economic resources to the value of 12,500 million pesetas. Another fundamental entry of the National Fund, are funds for the acquisition of scientific-technical infrastructure (27 billion pesetas). The last, and probably most important, entry corresponds to funds dedicated to the capitalisation of human resources, fundamental to ensure Spanish research potential in the long term. Investment was 57 billion pesetas dedicated to training and improvement actions.

Research Planning

The second objective of the National Plan for R&D is the planning of scientific and technological research. The National Plan sets out some general, performance guidelines for pluri-annual periods (three year periods up to the present) and tries to order activities in order to achieve this objective.

As mentioned in the section, Mechanisms for the Development of the National Plan, R&D planning is carried out through: National Programmes drawn up by the *CICYT*; Sectorial Programmes from ministerial departments or public organisations for action in specific topics; and lastly by regional government programmes, financed cooperatively and included in the National Plan due to national interest. The objective of planning is, apart from the correct orientation of the programmes, the avoidance of overlapping, duplication of efforts and administrative malfunctioning. This does not imply that planning must be absolutely rigid, but unclear definition of the programmes would be counterproductive. Programmes must be of a result-orientated nature and almost all should be determined from the beginning of the project; the majority of scientific projects should result in innovative products or processes of use in the Spanish socio-economic context.

The first National Plan for R&D was developed in the period 1988-1991 and included 28 programmes. Of these, two were sectorial and two were developed in the context of regional governments. The existence of overlapping and defects in administration were detected, bringing about the need to include these programmes in a more rigid and defined structure. Therefore the second phase of the National Plan (1992-1995) came about, incorporating the suggestions of organisations regarding planning. The number of programmes was reduced from 28 to 14, sectorial actions increased from two to four while the number of the regional government programmes remained the same.

Sectorial Programmes included in the National Plan have increased from 1 to 4 since 1988 (in 1988, the only Sectorial Programme was that of General Promotion of Knowledge of the Ministry of Education and Science). This means that the National Plan has started to include short term sectorial policies in its planning. This implies appropriate coordination and harmonisation with the other programmes of the Plan.

Another consequence of correct planning in R&D, is the multiplier effect of invested financial resources, both public and private. The reason for this is that the majority of the Plan's actions implies additional investments by the institution in which the activity is developed (or co-financing or general expense for the best development of the project). This multiplying effect has become of vital importance as far as resources dedicated to R&D are concerned, as it is estimated that, of total R&D financing by the National Plan, only one fifth has its origin in public administration.

Research Coordination

The third main objective of the National Plan is the coordination of efforts among the different agents involved in the development of projects. The fact that since 1988 a coordination mechanism has been systematised means that practically all public funds dedicated to R&D have been integrated into the National Plan, or act in accordance with the same instruments and methodology.

The effect of coordination can also be measured by the number of researchers (full-time) involved in actions related to the National Plan. In 1993, this figure corresponded to 53% of the total human resources in research in Spain. This percentage is even higher when referring only to the public sector.

Table 1-1: The National Plan: Mobilised Expenses

	National Fund	Sectorial Programmes	Total (Millions of pesetas)
National Plan	20,397	14,135	34,532
R&D Spending Mobilised	94,000	70,000	164,000

Source: SGPN, 1994

Another indicative parameter of the effects of coordination is the mobilising capacity of resources (already mentioned in section *Research Development*).

Only one fifth of the funds used in the financing of the National Plan activities come from the National Fund; the remainder originates in the different entities related to research projects. That is to say, the National Plan has been able to mobilise resources and capital toward the objectives and priorities of the Plan, although it is certain that, from the strictly economic point of view, the National Fund supposes little in relation to national R&D spending.

Maximum responsibility for coordination of the National Plan falls on the *CICYT*. A series of mechanisms have been created to develop this activity, facilitating communication and fluid exchanges among the agents involved.

In this context the National Plan created a network of Offices for the Transfer of Research Results (*Oficinas para la Transferencia de Resultados de Investigación - OTRI*), whose basic mission consists of establishing technology transfer links between public research centres, universities and enterprise. It should not be forgotten that the result of scientific production does not only depend on the scientific and technical quality of projects, not even when appropriately oriented by certain policy, but also interaction and adequate dissemination.

Initially, the *OTRI* network was set up exclusively as a link between public research centres and universities. Later, it was joined by different groups of researchers, and finally, under the suggestion of the *CICYT*, it was extended as an element of coordination between public and private research.

Basically, the network consists of a database of knowledge, infrastructure and R&D supply. It identifies and classifies the results obtained by research groups, evaluates their characteristics and disseminates information of interest to enterprise. This transfer can be direct or through other organisations nearer to enterprise. Another function is collaboration and participation in the preparation of research contracts, technical aid, consultancy, licenses of patents, etc. between the public research centres, universities and enterprise. They provide information on European R&D programmes, cooperating in the development of proposals to EU organisations, presenting and negotiating applications if required. Lastly, they collaborate in the exchange of research personnel between the universities and other entities where scientific projects are developed.

This dynamic structure is supported by the General Secretary of the National Plan through the Office of Technological Transfer (*Oficina de Transferencia Tecnológical - OTT*), created at the same time as the *OTRI*, with the purpose of coordinating and promoting the actions of the latter. Its functions are as follows:

- To compile information on the global technology offer (*DATRI* database) from universities, research associations, public research centres etc. and disseminate this information both nationally and internationally.
- To provide advice to *OTRI*'s regarding participation in EU programmes.
- To provide technical assistance to the *OTRI*'s as regards patents, projects agreed with companies, etc.
- To collaborate with the General Directorate for Scientific and Technological Research in the Research Personnel Exchange Programme between enterprise and public research centres.

- To collaborate with the *CDTI* in the dissemination of technology developed in public research centres, so that it can be implanted in industry.
- To organise training courses for the dissemination of knowledge.
- To establish dialogue among the different production agents, with the purpose of evaluating and detecting topics of applied research that must be developed.

Since 1988, the year of its creation, the *OTRI/OTT* network has encouraged links between enterprise and public research within the Spanish S&T system. The principal mechanism used to establish such bonds has been the development of mutually agreed projects (a National Plan project developed between a company and a public research centre) through the network.

In 1994, the number of offices making up the network was 74; universities (46), public research centres (10) and Enterprise Research Associations (18). Regarding *DATRI* (Database of Technology Supply), as an element of the *OTRI/OTT* network, the number of entries at the end of 1994 was 5,685. An important fact that should be mentioned, is that in recent years, direct access to this database via Internet has been made possible, thus promoting the search for Spanish partners by mainly European companies. Access to the network is also possible via: *CICYT, DGPYME, MEC-IBERTEXT, CSIC*.

Regarding the activities of industrial copyright, 143 patents were requested in Spain in 1994. Of these, 29 have been extended abroad promoting their legal protection beyond the national boundaries. 8 computer programmes were registered and 32 licenses signed. The data is reflected in the following table.

Table 1-2: Industrial Copyright Activities of the *OTRI/OTT* Network (1994)

	Patents Spain	Patents Foreign	Software	Licences
Universities	95	13	5	14
Public Research Organisations	42	15	3	18
Entrepreneurial Associations.	6	1	-	-
Total	143	29	8	32

Source: SGPN, 1995

The current tendency is that of reducing the number of contracts, but at the same time increasing the number dedicated to R&D activities and the funds administered by each one. In 1994, the number of contracts was 11,004 with total investment being 25,330 million pesetas. R&D related contracts were 2,436 in total. The remaining programmes concentrated on different areas (technical support contracts, cooperation agreements, training contracts and contracts for provision of services).

The administration of European projects through the *OTRI/OTT* network, according to financing requested, was 21.6% of the total in 1994. At present, the quantity assigned to Spanish projects is increasing substantially due to the initiation of the IV Framework Programme; it should be remembered that greater investment is made in R&D at the beginning of the Framework Programme and not at the end. In spite of this, given the lack of calls in specific programmes, the *OTRI/OTT* network has attempted to stimulate other EU actions (*INTAS, ALAMED, AVICENA, NEI*, etc.), as well as other types of action.

It can be stated that the *OTRI/OTT* network, apart from being a basic instrument of coordination, has encouraged R&D activities from two viewpoints. On the one hand, it has strengthened R&D in enterprise and on the other, has encouraged cooperative research between enterprise and public research centres.

Another instrument used by the National Plan to coordinate and evaluate public research are programmes within *PETRI*. The objective of these actions is to promote the relationship between the public research centres and enterprise; to ensure that, through aid and financing, research carried out in public organisations continues its development to the point where it can be applied industrially. These actions constitute an essential part of the National Plan for R&D, given that an S&T system would be of little use if industrial application did not form part of its development and did not materialise in improved competitiveness. The dissemination of results is carried out through the *OTRI* network. Here the interest of public administration can be seen, in giving a clearly socio-economic focus to research in public centres.

The National Agency of Evaluation and Forecasts also plays a key part in coordination. Its role of adviser provides it with knowledge on the programmes being carried out and the scope of their activities. Errors in coordination or overlapping, that may occur when introducing new research projects, can be avoided. Overlapping does not necessarily have to be of a technical character (duplication of effort); it may also be financial. The method used for project appraisal is that of peer review. Various independent experts intervene in the qualification of the project. The rigorous and independent operation of the *ANEP* has resulted in numerous entities, public and private, using this organisation to evaluate their scientific projects. Due to the broad diversity of its actions, this organisation has become a true coordination instrument of the Spanish S&T system. The objectives of scientific and technological forecasting, which should serve as a means to plan R&D, have not been developed to the necessary extent, perhaps due to insufficient structure.

The following basic element of coordination is the Sub-Programme of Exchange of Research Personnel which encourages the mobility of human research capital between enterprise and public research centres. The knowledge that can be acquired through such exchanges is fundamental in avoiding overlapping in the area of research and is an effective means of consolidating R&D departments in enterprise.

Lastly it should be pointed out that all coordination dedicated to obtaining competitive advantages in the Spanish S&T system would be impractical with-

Table 1-3: Contracts Negotiated by the *OTRI/OTT* Network- Areas of Action - 1994

	R&D	Technical Support	Cooperation Agreements	Training	Provision of Services	Total
Association	213	562	12	669	833	2,289
Public Research Organisation	408	127	104	29	133	801
University	1,815	1,706	506	860	3,657	7,914
Total	2,436	1,765	622	1,558	4,623	11,004

Source: SGPN, 1995

Table 1-4: Contracts Negotiated by the OTRI/OTT Network According to areas of action - 1994 (Millions of pesetas)

	R&D	Technical Support	Cooperation Agreements	Training	Provision of Services	Total
Association	2,887	484	573	522	572	5,040
Public research Organisation	2,691	951	256	304	192	4,396
University	7,940	2,702	1,348	2,694	1,207	15,893
Total	13,518	4,139	2,178	3,521	1,973	25,330

Source: SGPN, 1995

Table 1-5: Number of Contracts Negotiated by the *OTRI/OTT* Network According to Type of Contracting Entity - 1994

	Administration	Enterprise	Others	Total
Association	180	2,099	10	2,289
Public Research Organisation	205	531	65	801
University	1,647	5,626	641	7,914
Total	2,032	8,256	716	11,004

Source: SGPN, 1995

Table 1-6: Value of Contracts Negotiated by the *OTRI/OTT* Network According to Type of Contracting Entity (1994). (Millons of Pesetas)

	Administration	Enterprise	Others	Total
Association	1,650	3,247	142	5,040
Public Research Organisation	1,402	2,603	390	4,396
University	6,324	7,050	2,518	15,893
Total	9,377	12,901	3,051	25,330

Source: SGPN, 1995

out being linked to R&D policies, especially those developed within the framework of the European Union. The basic instrument of the EU toward achieving its objectives, is the Fourth R&D Framework Programme in which the high-priority areas of research are included as well as financing necessary to carry out such lines. The objectives of the programmes are of common interest for all involved countries; the coordination factor becomes vital. The organisation in charge of supervising link actions is the Spanish Link Centre (VALUE), recently set up and co-administered by the *SGPN* and the *CDTI*. Its mission is the diffusion of technology acquired from the different framework programmes in the web of Spanish industry, well as the diffusion of that acquired in the Spanish public centres, among European companies.

NATIONAL PROGRAMMES

This section describes the programmes that make up the National Plan for R&D and the funds assigned in recent years.

As has already been seen, the main elements of the National Plan are the National Programmes. These programmes have been centred on 4 high-priority areas since 1988. These areas are:

- Production and Communication Technology.
- Quality of Life and Natural Resources.
- Social, Economic and Cultural studies.
- Horizontal and Special Programmes.

The National Plan also takes in sectorial initiatives and regional government programmes. The following tables shows such programmes. The first gives details of those developed in the II National Plan for R&D (1992-1995), while the second outlines the programmes of the recent III National Plan (1996-1999).

Table 1-7: II National Plan for R&D (1992-1995)

NATIONAL PROGRAMMES:

PRODUCTION AND COMMUNICATION TECHNOLOGIES:
Advanced production technologies.
Information & Communication Technology
Materials.
Space research

QUALITY OF LIFE AND NATURAL RESOURCES:
Biotechnology.
Agricultural sciences.
Environment and Natural Resources.
Health and Pharmacy.
Food Technology.

SOCIAL, ECONOMIC AND CULTURAL STUDIES:
Social, Economic and Cultural studies.

HORIZONTAL AND SPECIAL PROGRAMMES:

Training of Research Personnel.
Research in the Antarctic.
High Energy Physics.
Information for Scientific Research and Technological Development.

REGIONAL GOVERNMENT PROGRAMMES:

Fine chemistry (Catalonia).

SECTORAL PROGRAMMES:

General promotion of Knowledge (Ministry of Education and Science).
Teacher Training & Improvement of Research Personnel (Ministry of Education and Science).
Agricultural and Food R&D (Ministry of Agriculture, Fisheries and Food).

Source: SGPN, 1995

**Table 1-8: III National Plan for R&D
(1996-1999)**

NATIONAL PROGRAMME:

PRODUCTION & COMMUNICATION TECHNOLOGIES:
Advanced production technologies.
Information and Communication Technologies.
Materials.
Space research.
Telematic applications and services.
Chemical Process Technologies.

QUALITY OF LIFE AND NATURAL RESOURCES:
Biotechnology.
Health.
Agricultural R&D.
Environmental R&D.
R&D on the Climate.
Hydraulic Resources.
Science and Marine Technologies.
Research in the Antarctic.

SOCIAL, ECONOMIC AND CULTURAL STUDIES:
Social, Economic and Cultural studies.

HORIZONTAL AND SPECIAL PROGRAMMES:

Training of Research of Personnel.
Strengthening of the Structure of the Scientific-Technology-Industry System (PACTI).
High Energy Physics.
Information for Scientific Research and Technological Development.

REGIONAL GOVERNMENT PROGRAMMES

Fine chemistry (Catalonia).

SECTORIAL PROGRAMMES:

General Promotion of Knowledge (Ministry of Education and Science).
Teacher Training and Improvement of Research Personnel (Ministry of Education and Science).
Agricultural and Food R&D (Ministry of Agriculture, Fisheries and Food).
Studies on Women and Gender (Ministry of Social Affairs).

Source: SGPN, 1995

The following figure shows the distribution of funds by areas in National Plan programmes in 1994:

Table 1-9: Balance of Administration of the National Plan for R&D - Millions of pesetas

Programmes	Training of Research Personnel	Economic Commitment to Projects	Projects, Infrastructure & Special	Mutually Agreed Projects	Other Spending + *PETRI*	Total
Agricultural sciences	900.000	152.064	595.100	270.000	20.451	1,937.615
Food Technology	100.000	108.867	589.423	360.500	14.105	1,172.895
Environment	220.000	202.266	1,248.455	200.000	5.208	1,875.929
Biotechnology	400.000	208.136	898.015	265.000	37.053	1,808.204
Health & Pharmacy	200.000	217.106	709.087	360.000	24.085	1,510.278
Materials	950.000	307.366	1,732.221	1,080.000	27.050	4,096.637
Advanced Production Technology	90.000	169.472	619.214	795.000	10.605	1,684.291
Information & Communication Technology.	200.000	311.402	2,157.182	754.000	14.233	3,436.817
Space Research	100.000	64.262	201.680	145.000	-	510.942
Fine chemistry (1)	-	32.945	230.000	-	-	230.000
Social studies	115.000	192.708	132.797	-	-	280.742
High Energy Physics	50.000	18.092	193.535	-	500.000	936.243
Research in the Antarctic	10.000	-	429.102	-	119.000	576.194
Scientific Information & Technological Development	15.000	-	230.279	-	-	245.279
Training of Research Personnel. - Horizontal actions	650.000	-	-	-	-	650.000
Actions, Scientific Politics & OTT					118.264	118.246
New Actions					704.700	704.700
Total	4,000.000	1,984.686	10,166.090	4,229.500	1,594.736	21,775.012

(1) The figure indicated corresponds to the transfer made to the Generalidad de Cataluña as a contribution by the CICYT. Additionally, another 700 million pesetas from the European Social Fund have been invested.

Source: SGPN, 1994

Figure 1-6: National Plan Programmes: Breakdown of funds by scientific-technical areas. (1994)

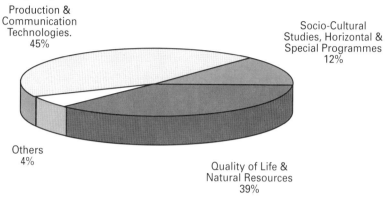

Total: 21,775 Millions of pesetas.

Source: SGPN, 1994

Figure 1-7: Breakdown of National Funds by Activity (1994)

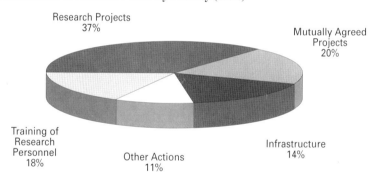

Source: SGPN, 1994

The National Fund, as a budgetary instrument of the National Plan, received 21,775 million pesetas in 1994. The global balance of administration of this fund may be seen in Table 1-9 by programme and activity, showing financing for actions developed in that year. A column has also been added with the funds corresponding to economic commitments acquired in previous years with research and integrated projects.

As can be observed, both funds dedicated to infrastructure and those dedicated to research projects constitute the most important budgetary headings. Again, the mobilising effect of National Plan resources must be emphasised. Perhaps the figure of 21,775 million pesetas dedicated to the development of R&D in 1994 is not excessive in comparison with total R&D expenditure in Spain, but it priority areas of national interest.

In the area of Production and Communication Technology (Materials; Advanced Production Technologies; Aerospace Research; Information and Communication Technologies), almost half of the funds were dedicated to the realisation of research projects and the provision of appropriate infrastructure. Within these, the majority of funds were dedicated to Advanced Production Technologies and Information and Communication Technologies. For the training of research personnel alone, 10% of the total was used. The presence of enterprise in these fields of research becomes clear in the fact that more than 30% of funds is dedicated to agreed projects, due mainly to the disseminating nature of the technologies included in this area. That of Materials is particularly noteworthy in relation to the other agreed projects.

If investment is classified by scientific-technical objectives, the Materials Programme represents 42% of investment; Information and Communication Technologies, 35%; Space Research 5%; and Advanced Production Technologies, 15%.

After the areas of Production and Communications Technologies, the programmes of Quality of Life and Natural Resources (Agricultural Sciences, Food Technology, Environment and Natural Resources, Biotechnology, Health and Pharmacy) are those which absorb greater funds. The breakdown of these programmes as regards the budget is quite similar. In 1994: 23% of funds was dedicated to Agricultural Sciences, 14% to Food Technology, 22% to Environment and Natural Resources, 21% to Biotechnology and 18% to Health and Pharmacy. In comparison to other research areas, here the level of funds dedicated to training reaches significant values, Agricultural Science with 21% showing the highest level of effort. Mutually agreed projects on the other hand, are not as significant as the Programme of Production Technologies, where in the areas of Health, Agriculture, Environment, etc. the presence of enterprise is much lower, and it is public administration which must carry out most of the research. In the area of investment in infrastructure, funds destined for Environment and Natural Resources have substantially increased in recent years, reaching 40% of the total figure dedicated to this type of action.

Horizontal and Special Programmes, due to their peculiar characteristics, cover 2.4% of the budget of the National Fund. Worth special mention is the High Energy Physics Programme, associated to Spanish participation in CERN, to which 40% of the total funds of these programmes has been dedicated. The rest, Research in the Antarctic (through the Spanish Base, *Juan Carlos I*, in Antarctica and the oceanographic ship *Hespérides*), the Information for the Scientific and Technological Research programme and the programme for the Training of Research Personnel receives smaller quantities.

In the following figure, breakdown is according to high-priority areas.
As can be observed, the funds dedicated to the different areas do not show similar distribution. This indicates that there exist preferred areas of action within the planned structure of the National Plan. The area of production and communica

Figure 1-8: Programmes of the National Plan: Distribution by High Priority Areas (1992-1995)

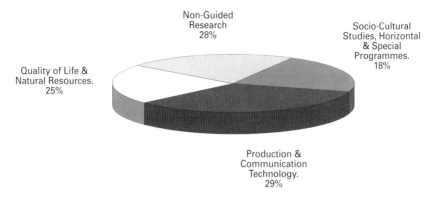

Source: SGPN, 1995

tion technologies monopolises 29% of the funds, due to the impulse that the Government wishes to apply to the industrial sector, a key factor for the convergence of Spain with the rest of Europe.

The other fundamental entry shown in the figure, forming part of the various programmes, is that of non-guided research, which represents 30% of the total.

Coordination of National Programmes

National Programmes are a primary coordination principle when orienting scientific projects toward common objectives in line with the national interests determined by the Government. These objectives also include sectorial or regional initiatives of interest. The mechanisms and institutions engaged in coordination activities have already been identified. The most important coordination actions carried out to date will now be outlined, enabling the reader to form an idea of their level and development. Coordination must be carried out at different levels keeping within the scope of the programmes. Reference is made only to coordination of the Sectorial Programs, Regional Programmes, territorial coordination and coordination with non-European international programmes. Programmes carried out within the European Union will be described and analysed separately.

Coordination of Sectorial Programmes

Sectorial programmes are a fundamental means of integrating R&D initiatives of the different ministries into the National Plan. This ensures appropriate coordination with other programmes carried out. To date, such harmonisation has only been partly achieved.

Regarding the Sectorial Programme of General Promotion of Knowledge, a high degree of coordination has been achieved between the Ministry of

Education and Science and the other programmes which constitute the National Plan. Since 1988, the date of initiation of this programme, coordination of activities with other National Plan activities has been efficiently carried out through a flexible mechanism of initiative transfer, regarding both projects and infrastructure. The fact that evaluation and follow up of the programme is carried out by the National Agency for Evaluation and Forecasts has greatly benefited coordination tasks.

The second Sectorial Programme included in the second phase of the National Plan, was Training and Improvement of Research Personnel. Given that a National Training Programme for Research Personnel exists, coordination between both is of great importance. A method of unified administration was developed to ensure coherent criteria, avoiding the undesirable effect of overlapping and duplication of effort. Advances in coordination between both programmes has undergone important development in recent years.

In comparison with these two programmes, where harmonisation of actions is at least efficient, either there is a lack of coordination or it is not being conducted with appropriate linkages in the others.

With respect to the activities of the Ministry of Agriculture, Fisheries and Food, the National Plan for Agricultural Research, administered by the National Institute for Agricultural and Food Research and Technology, was established in 1986, date of the enactment of the Science Law. This programme was not effectively integrated into the National Plan until 1991, becoming a Sectorial Plan of Agriculture and Food Research of the Ministry. The programme objective is to direct short term support by the Ministry of Agriculture to regional governments in agricultural matters. In order to fulfil such an objective, it was necessary to maintain active a specific support programme beyond the framework of the National Plan. Coordination between both is effective at present, thanks to a cross-cutting administration system in which representatives and involved agents participate in the management of both programmes.

Although coordination with the Ministry of Health and Consumption regarding R&D activities is in the initial stages of development, progress to date has been significant. Principal coordination activities have been carried out through financing provided by the Health Research Fund (*Fondo de Investigaciones Sanitarias - FIS*). The most important link to be developed is that formed by the National Health and Pharmacy Programme, General Promotion of Knowledge and the *FIS*, facilitating the flow of information, harmonising administration and evaluation systems, and tying together the different objectives. The *CICYT* included the *FIS* Programme, as a Sectorial Programme, within the National Plan in 1994 in order to achieve the above.

Numerous coordination actions still have to be carried out at the sectorial level. The expansion of such activities should be effectively implemented to serve as a driving force for the inclusion of ministerial R&D initiatives in global harmonisation. Activities of wide sectors of public administration are still on the fringe of the National Plan. Should this be corrected, the Plan will become appropriate support for the main areas of sectorial performance by public administration.

Coordination of Regional Government Programmes

Great disequilibrium, both economic and regarding R&D, exists among the different Spanish regions. Due to this, various regional governments have initiated their own plans for research development. The weight of this financing has slowly grown in the context of the National Fund. Initially, this fund was not designed to reduce regional imbalance, but rather as an eminently, competitive instrument. The National Plan, however, can act as a catalyst of regional cohesion and for this reason, the *CICYT* has developed various actions with this aim. The main one is that which allows regional activities regarding R&D to be integrated into the National Plan, should they demonstrate that they are of special interest to government proposed objectives for the country as a whole. In accordance with Article 6 of the Science Law, this type of regional programmes included in the National Plan acquire the same range as National or Sectorial Programmes.

Since the initiation of the National Plan (1988), two regional programmes have been included:

- Fine Chemistry (*Generalidad de Cataluña*, 1989)
- New Technologies for the Modernization of Traditional Industry. (*Generalidad Valenciana*, 1989)

The objective of the Fine Chemistry Programme is the development of R&D in the Catalan chemical industry. The programme was set up in 1989 by the regional government and is administrated by the Interdepartmental Commission for Research and Technological Innovation (*Comisión Interdepartamental de Investigación y Innovación Tecnológica - CIRIT*). In the same year, the programme was included in the structure of the National Plan.

Apart from the promotion of R&D in the chemical sector, this regional government programme finances particular actions in public research centres, co-finances all types of mutually agreed projects between public research centres and enterprise and the improvement of existing infrastructure.

Another prominent action is the creation of a Centre for Process Development, with investment approaching 400 million pesetas, the administration of which corresponds by public tender to the Testing and Research Laboratory of the *Generalidad de Cataluña*.

The programme, New Technologies for the Modernisation of Traditional Industry, was included in the National Plan in 1989 at the proposal of the *Generalidad Valenciana*. As suggested by the title, the programme promotes R&D actions in traditional sectors of the region. The most deeply rooted industry in the region of Valencia has clearly declined in recent years, due to lack of decisive actions in the area of R&D and managerial innovation. The action started out with the objective of solving this set of "bad structures." The primordial objective has been the orientation of this industrial sector toward high value added products, stressing product differentiation and emphasising a clear commitment to quality. The means of developing this objective has been the introduction of significant improvements in administration and production processes.

From the date of the project's initiation (1989) up to 1991, over 40 research projects were carried out. Investment was greater than 700 million pesetas. The programme was terminated in 1991 at the request of the *CICYT*, as the objectives were considered as having been fulfilled.

Other National Coordination Actions

So far, only coordination of the more usual action mechanisms of the National Plan, i.e. Research Projects, Sectorial Plans, and Regional Government Programmes have been dealt with. In addition, other types of sectorial coordination exist, whose mission is to increase entrepreneurial competitiveness through the introduction and dissemination of actions of technological and innovative development. Evidently this type of action, owing to the fact that it is applicable, must be developed in the short or mid-term. Such actions are executed in close collaboration with the usual agents in the area of research; public research centres, universities, various levels of public administration and enterprise.

The most usual mechanism for the execution of this type of action are integrated projects. These are basically orientated toward R&D in concrete topics and demand the mobilisation of a large quantity of resources and the integration of an ample range of technologies contributed to by diverse elements of the Spanish S&T system. Concrete examples of these projects would be those integrated in the Aerospace Research Programme (SOHO and MINISAT) and in Information and Communication Technologies (*PLANBA*).

A series of actions, such as the *MIDAS* (*Movilización de la Investigación, el Desarrollo y las Aplicaciones de Superconductores* - Mobilisation of Research, Development and Superconductor Application) programme, have been drawn up to develop joint programmes with the entrepreneurial sector. Companies of the electricity sector, such as *Unión Eléctrica Española* or *Red Eléctrica Española*, and the *CICYT* participate in the project. Another habitual mechanism are special actions, among which the programmes *GAME*, *PACE* and *PASO* can be highlighted. These are initiatives of the European Commission, but coordination at national level is carried out by the *CICYT*. Lastly, the Ministry for Industry and Energy and the *CICYT* are responsible for coordination of Spanish R&D in areas related to microelectronics and software.

The third field of action is the optimisation of the use of large scientific centres, both by enterprise and public administration. A high-priority objective is also greater integration of international programmes in their normal activities. The centres referred to are the same as those mentioned in the section on Public Research Centres (*IAC*, *BIO Hesperides*, etc.). Various committees also exist to coordinate Spanish activity in large international research centres such as CERN, ESA etc.

The National Agency of Evaluation and Forecasts has served as a basic element for the coordination of many enterprise programmes. It is quite normal that the R&D departments of specific companies go to this Agency seeking advice on how to carry out their research activities. This is due to the independent and rigorous operation of this Institution, which has been transformed into a basic coordination element of the Spanish S&T system.

EUROPEAN R&D FUNDS

European Regional Development Fund

The basic mission of these funds is inter-regional cohesion in the European Union. The resulting tasks are carried out with this in mind. Since 1989, the development of various actions related to R&D in lesser favoured regions of the EU has also been included. The coordination and distribution, according to priority areas throughout Objective 1 and 2 regions, of the different R&D resources provided by these funds has been a fundamental mission of the *CICYT* since the outset. That is to say, it manages all the resources destined for central government. The *CICYT* is also responsible for coordinating the different regional programmes financed by these funds and co-administered by the regional governments.

Within this framework, the *CICYT* designed and managed the administration of the Operative Programme for R&D Infrastructure for Objective 1 regions, in the period 1991-1993. It also planned the actions to be developed in other more favoured regions of Objective 2. This planning was executed in such a way as to maximise coordination between central public administration and the programmes to be developed with regional governments, through their regional programmes, on a co-financing basis.

In 1991, the STRIDE initiative began. Its basic objective is the promotion of enterprise competitiveness at all levels, an objective which is not limited to increasing enterprise development inside the country, but also covers international promotion of companies. STRIDE's actions are basically aimed at SMEs, and improvements in competitiveness are achieved through innovative projects and improved organisation.

The tasks of coordination are fundamental when administrating ERDF funding, and given the existence of the National Plan, the appearance of a new source of financing implies greater organisational effort and coordination.

On the whole, these actions represented an investment of European funds in R&D infrastructure to the value of 25 billion pesetas, concentrated in lesser developed regions. This quantity represents approximately 55% of the total investment in R&D infrastructure by universities in these regions in the period 1991-1993, giving an idea of the magnitude and importance of this source of financing. A new phase of the application of the ERDF began in 1994. The objectives were re-outlined, dedicating greater quantity of the funds to actions of technological improvement and industrial development. This was due to the fact

Table 1-10: Operative ERDF Programmes (1989-1993): Objective 1 Regions.
(Millions of pesetas)

	Total investment	ERDF Aid
Andalusia	6,155	3,692
Asturias	938,3	563
Canary Islands	2,213	1,328
Castile - La Mancha	1463	877
Castile and León	2,673	1,603
Extremadura	949	569
Galicia	2,050	1,230
Murcia	983	589
Valencia	4,692	2,815
Total	22,117	13,270

Source: SGPN

Table 1-11: Operative ERDF Programmes (1989-1993): Objective 2 Regions.
(Millions of pesetas)

	Total investment	ERDF Aid
Aragon	1,261	630
Cantabria	305	152
Catalonia	5,434	2,717
Basque Country	1,300	650
Madrid	4,581	2,290
Navarre	560	280
Total	13,441	6,720

Source: SGPN, 1995

Table 1-12: Comparison of Structural Funds (1989-93/1994-99).
Total investment (Millions of pesetas)

	1989-1993	1994-1999
Objective 1*	28,868	74,309
Objective 2*	14,632	14,296
REGIS **	0	3,529
INTERREG II	0	3,001
Objective 5B	0	680
Total	43,500	95,815

* *STRIDE is included for Objectives 1 and 2*
** *Data regarding Objective 2, GOVERN II and INTERREG II is provisional.*

Source: SGPN, 1995

that strengthening the productive sector of a region automatically and directly improves its socio-economic level. In global terms, the ERDF represents 1.2% of the national total invested in R&D infrastructure. The forecast rate of growth of this investment level is situated at a constant 2.8% for the period 1994-1999.

In the last year, the Operative Pluri-regional Programme for R&D Infrastructure in Objective 1 regions has been passed. Application began in 1994 and is programmed to end in 1999. The *CICYT* collaborated significantly in coordination between the national objectives of the National Plan and the individual concerns of interested regional governments. In the redesign of objectives and priorities, the experience acquired in the previous phase of execution was fundamental, leading to the establishment of much more ambitious goals. The total level of investment foreseen for this period is 53 billion pesetas.

Within the actions foreseen in this new framework, a specific programme was included for the training of research personnel, diversifying the offer of actions that initially were dedicated only to the creation of new R&D infrastructure or improvement of the existing one. This new financing will suppose investment of an additional 17 billion pesetas for Objective 1 regions.

Tables from 1-10 to 1-13 show the breakdown of the ERDF by autonomous region. Table 1-12 shows the figures forecast for the period 1994-1999, which as can be seen increased regarding those of the previous period indicated.

Table 1-13: Regional Distribution of Investment. Objective 1. 1994-1999. (Mill. of Pesetas)

	Total investment	ERDF Aid
Andalusia	14,123	10,239
Asturias	5,256	3,679
Canary Islands	6,807	5,105
Cantabria	3,540	2,478
Castile - La Mancha	6,599	4,619
Castile and León	10,049	7,034
Extremadura	3,890	2,917
Galicia	5,270	3,821
Murcia	4,850	3,395
Valencia	9,421	6,594
No specific region(*)	4,501	3,150
Total	74,309	53,036

* *Agreed projects.*

Source: SGPN, 1995

INTERNATIONAL FRAMEWORK

European Union Framework Programme

As already mentioned, the framework of actions regarding R&D is drawn up in the European Single Act and the Framework Programme is set out as the basic instrument for the development, planning and coordination of scientific policy within the European Union. This Programme includes the basic, high-priority areas of research, contained in specific programmes which are designed to satisfy needs, both on an individual country basis and for the group as a whole. In addition to the programmes themselves, the resources assigned to them are also established and the actions through which they will be developed. The importance of the Framework Programme must be underlined, not only in its context of a source of financing, which is enormous, but also as an element of the driving-force of the Spanish S&T system. It could even be stated that the Framework Programme constitutes an important effort of regional cohesion, although not its fundamental objective, serving as an element of influence on the various Spanish agents, especially in lesser developed areas.

To get an overall view of the above, it is sufficient to show the level of financing granted by the European Commission to the Spanish programmes developed within the Framework Programmes. Data is included regarding participation in the Framework Programmes I, II and III, but not on IV, due to the fact that at present, data is not all available. The following figure shows the credits, in millions of ECUs, granted to research in Spain.

Figure 1-9: Credit Granted to Research Framework Programme (Millions of ECUs)

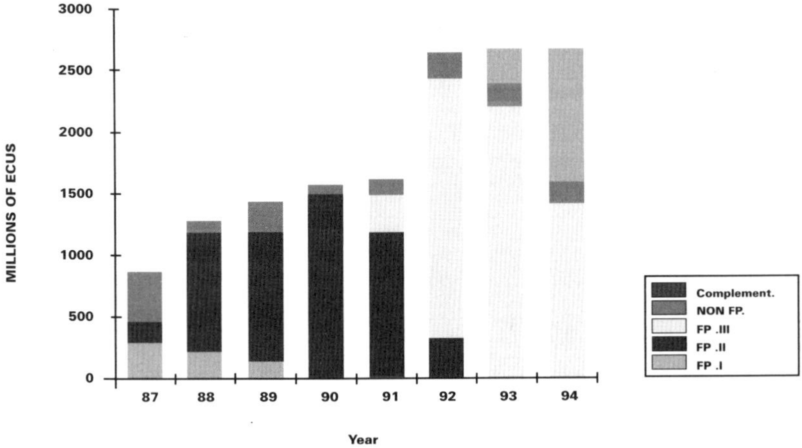

Source: SGPN, 1995

As for Spanish participation in the III Framework Programme in comparison with the participation of the other EU countries, Figure 1-10 shows the percentage of financing regarding the total dedicated to scientific programmes.

Figure 1-10: European Participation in the Financing of the III R&D Framework Programme

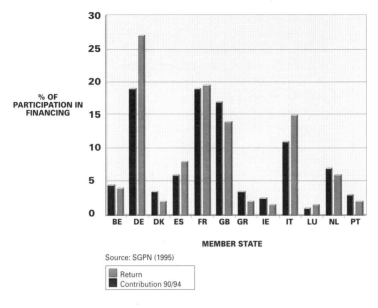

Source: SGPN, 1995

Spanish participation, in spite of not being excessively high if compared with countries such as France or Germany, is still significant with 8.3% of the total. This participation has been conditioned by diverse factors, of which the following should be highlighted: the relative position of Spain regarding the most advanced countries of this context, lower dedication of human resources to this type of programme and the high level of competitiveness in the Framework Programme. The percentage of returned funds (i.e. the quantity returned to the country in the form of both company and public administration contracts) of total financing is 6.3%. The relation between financed and returned can be classified as acceptable if compared with other countries of this background.

It is also necessary to highlight greater versatility in the training of research personnel, exchange of knowledge due to the ease of research personnel exchange, and the wide margin of SME activity due to their ability to participate in programmes together with other European companies.

The percentage of Spanish collaboration in the total of projects carried out in the European Union can also be considered satisfactory. Spain has participated in more than 24% of the programmes approved by the European Commission which have a marked industrial character. It is also important to mention the fact that the level of SME participation in these projects is very high, both in technological innovation and industrial development. This is of vital importance given that the small company serves as a driving-force in the national socio-economic environment. In the long term, Spanish presence in EU projects faci-

Figure 1-11: III Framework Programme: Participation by Countries.

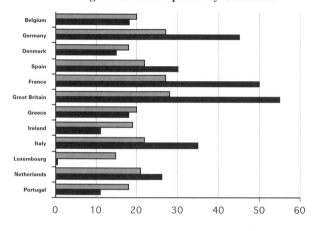

Source: SGPN (1995)

■ % Projects/Total Projects
■ % Proposals/Total Proposals

Source: SGPN, 1995

Table 1-14: III Framework Programme. Spanish Participation

Programmes	Total Financing (Thousands of ECUs)	Financing - Spain (Thousands of ECUs)	%	Total Projects	Projects With Spanish Participation	%
Information Technology	1,323,700	82,300	6.6	689	183	26.6
Communication Technology	523,000	26,600	5.2	189	71	37.6
Telematic Systems	326,000	18,100	5.6	292	98	33.6
Industrial & Material Technology	751,200	50,500	7.2	528	143	27.1
Measures and Testing	37,700	3,300	8.7	112	67	59.8
Environment	293,300	17,900	6.1	564	143	25.4
Sciences & Marine Technologies	100,400	3,500	3.5	84	22	26.2
Biotechnology	147,100	11,400	7.7	183	67	36.6
Agriculture & Agro-industry	340,900	31,900	9.4	420	179	42.6
Biomedicine & Health	128,300	6,200	4.8	409	174	42.5
Coop. with Developing Countries	99,100	3,900	3.9	262	48	18.3
Non Nuclear energy	219,400	11,600	5.3	458	90	19.7
Nuclear Fission Safety						
Controlled Thermonuclear Fusion	39,405	10,750	27.2			
Human Capital & Mobility	508,600	36,700	7.2	3.266	730	22.4
Total	4,798,700	303,900	6.3	7.456	2,015	27

Source: SGPN, 1995

litates the creation of permanent bonds with the participant societies at the European level, facilitating permanent transfer of technology and gradual enrichment of the Spanish S&T system.

In Figure 1-11, the level of Spanish participation in EU projects is shown, and clearly indicates the high level of participation mentioned above. As can be observed in the previous table, Spanish participation in programmes of a marked industrial nature is outstanding in that it received more than 70% of the funds and made up more than 75% of Spanish returns. These results are a clear reflection of Spanish business interest in the development possibilities made available by participation in the Framework Programmes, especially in sectors of state of the art technology. This situation is shown by the fact that the programmes of Communications Technology, Information Technology, Industrial and Material Technology are those programmes with a higher level of Spanish participation. Therefore resources assigned are much greater.

Concentrating on Spain (Table 1-14), the following fact to outline is Spanish participation in the projects constituting the R&D Framework Programme of the EU. The percentage granted to Spain is 8.3%, and participation in programmes is over 24%.

To summarise the data, Table 1-15 shows Spanish participation in the III Framework Programme. Information regarding participation in the second Programme has also been included, allowing the observation of the increase of effort by the Spanish authorities in adapting European initiatives as regards R&D to the national interests of the Spanish S&T system.

The table shows that the Spanish position with respect to R&D is notably inferior to that which corresponds to the weight of its economy. The Spanish contribution is made fundamentally in function of this weight. It is not surprising therefore, that the global balance in the Framework Programme (contribution compared to returns) is negative. This situation is confirmed by the fact that Spanish researchers represent only 6.6% of the EU total, the Spanish population being much larger percentage wise. If reference is made only to researchers involved in business activities, the true driving-force of the S&T system, the percentage is substantially reduced.

The percentage of projects of the EU with Spanish participation and participation of other EU countries is shown in Figure 1-12. It can be observed that France, Germany, Great Britain and Italy are preferred partners. The reason for this is simple. Spanish research groups prefer the collaboration of countries with long research traditions, given that project execution is more effective and technology transfer, greater.

Collaboration with Southern European countries, Greece and Portugal, is also prominent, although to a lesser extent.

Concentrating on Spain, the following fact to outline is Spanish participation in the projects constituting the R&D Framework Programme of the EU. The percentage granted to Spain is 8.3%, and participation in programmes is over 24%

Table 1-15: Spanish Participation in EU Framework Programmes

	II Framework Programme	III Framework Programme
Total financing (thousands of ECUs)	3,575,250	4,798,700
Returns Spain (thousands of ECUs)	199,000	303,900
%	5.5	6.3
Total Projects	4,487	7,456
Projects with Spanish participation	983	2,015
%	21.9	27.0

Source: SGPN, 1995

Figure 1-12: Percentage of Spanish Projects With Partners from other Member States

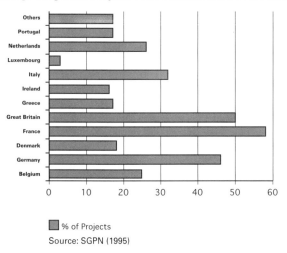

Source: SGPN, 1995

Breakdown by project leader according to country, shows Spain's role in the different projects. Should a country participate as leader in a high percentage of projects, international recognition of researchers and centres can be affirmed. This is not so in the case of Spain. Spain acts as leader and organiser in only 17% of the projects in which it participates, lagging behind France and the United Kingdom with superior percentages (Figure 1-13). This indicates that Spanish participants are not regarded as prestigious within the EU. In spite of everything, and contrary to what may appear, the average ratings obtained in the projects led by Spain is around the EU average. It can be stated, that the level of Spanish scientific competence is not lower than that of the other Member States.

Figure 1-13: Distribution by Leaders: Projects with Spanish Participation

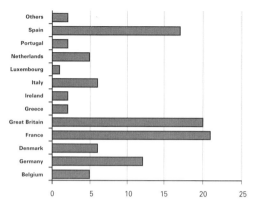

Source: SGPN, 1995

As for Spanish participation according to the type of entities involved in the development of projects, companies monopolise 40% of the total financed, followed by public research bodies with 31%. The remainder is shared between universities with 28% and research associations with 3%. As for the percentage of projects approved regarding the total, the results are unlike those of financing. Spanish companies are responsible for the development of 20.5% of projects, while public research centres are responsible for 34.4%, universities 43.6% and research associations 1.5%. This data mainly indicates the greater importance of programmes developed in companies and the greater assignment of funds needed to develop their activities as regards R&D. The importance of these results must not be disassociated from the fact that the percentage of Spanish researchers in industrial centres does not even reach 30%, from which one can deduce the driving effect of EU programmes in national industry. Another fact to be kept in mind, is the pre-competitive character of the projects developed in the framework of the Union, indicating the interest of companies in these areas of research.

In the future, and with the initiation of the IV Framework Programme, it is expected that participation by enterprise will be significantly strengthened.

The data shown in Table 1-16 reflects the following; the number of projects developed under Framework III, the number of entities taking part (keeping in mind that an organisation can participate in more than one research project), the total financing in millions of ECUs received and distribution by region.

As can be observed, a clear relation exists between the number of groups involved and the financing available. It is evident that the European Union assigns special value, not only to transnational technology transfer, but also to that within the national boundaries. This correlation is also supported by the fact that in those regions where the human capital dedicated to research is higher, the level of research development, and therefore investment, will also be higher. Both

Figure 1-14: III Framework Programme: Spanish Participation by Type of Entity. Projects Approved.

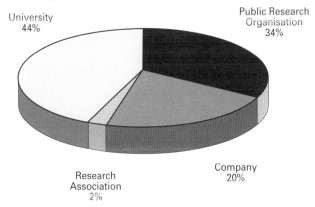

Source: SGPN, 1995

Figure 1-15: III Framework Programme. Spanish participation by type of entity. Financing.

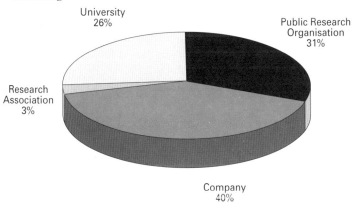

Source: SGPN, 1995

the Autonomous Region of Catalonia and that of Madrid monopolises more than half of scientific production, both in the number of projects and financing obtained. Finally, the returns obtained by regions with lower technological development are also more reduced. In practice, this unbalanced situation in the context of the Framework Programmes is somewhat eased by the assignment of funds to R&D through the European Structural Funds. As pointed out previously, the objective of the latter was inter-territorial cohesion, rather than scientific and technological development.

The IV Framework R&D Programme began its approval of research projects in 1994. Proposals for improvement, based on the experience acquired by the Member States throughout the development of the two previous programmes,

Table 1-16: Spanish Participation in the Framework Programme: Distribution by autonomous regions - %

	N° of Projects	N° of Groups	Total Financing
Andalusia	8.9	8.5	7
Aragon	2.9	2.7	1.8
Asturias	1.5	1.4	1.6
Balearic Islands	1.4	1.4	1
Basque Country	6.2	6.4	9.2
Canary Islands	1.7	1.6	0.7
Cantabria	1.4	1.2	1.2
Castile and León	2.5	2.4	1.6
Castile-La Mancha	0.4	0.4	0.3
Catalonia	23.5	23.3	17.8
Extremadura	0.4	0.4	0.2
Galicia	2.1	2.3	1.7
Madrid	36.8	38.8	48.6
Murcia	1.4	1.3	1
Navarre	0.9	0.9	0.7
Rioja	0.1	0.1	0
Valencia	7.5	7	5.4

Source: SGPN, 1995

were included. The activity of the *CICYT* has been constant during the last four years regarding the negotiation of programmes and the mobilisation of Spanish groups to participate in European consortia, and in defence of Spanish proposals and negotiation of corresponding contracts. In recent years, work relating to the coordination and execution of the III Framework Programme has overlapped with the administration of the IV Programme and approval of corresponding programmes.

In relation to the IV Framework Programme, important changes have been introduced in relation to the III Programme. First, a spectacular increase has occurred in the assignment of funds to research projects. This is due to the understanding by the Member States of the importance of having an efficient S&T system on an individual basis, that can contribute the necessary technological and innovative development to maintain Europe at the highest level in world competitiveness. Secondly, two new research areas have been introduced (Transport and Socio-Economic Research), while Non-Nuclear Energy and Life Sciences have been granted higher quantities of resources. Dissemination activities on research results and the training and mobility of researchers have been reinforced. Lastly, an action relating to the development of cooperation with third countries (developing countries and Eastern European countries) and with international organisations has been included.

The *CICYT* made a special effort in defence of the Programme for Dissemination and Evaluation of Research Results and Training and Mobility of

Table 1-17: Budgets of the III & IV Framework Programme

	III Framework Programme		IV Framework Programme	
	Million ECUs	%	Million ECUs	%
1st Activity				
Information & Communication Technology	2,516	38.1	3,045	27.68
Information Technologies	1,532	23.2	1,932	15.71
Communication Technologies	554	8.4	630	5.12
Telematic Systems	430	6.5	843	6.85
INDUSTRIAL & MATERIALS TECHNOLOGY	1,007	15.3	1,995	16.22
Industrial & Materials Technology	848	12.9	1,707	13.88
Measures and Testing	159	2.4	288	2.34
ENVIRONMENT	587	8.9	1,080	8.78
Environment	469	7.1	852	6.93
Marine S&T	118	1.8	228	1.85
LIFE S&T	714	10.8	1,572	12.78
Biotechnology	186	2.8	552	4.49
Agriculture	377	5.7	684	5.56
Biomedicine and Health	151	2.3	336	2.73
ENERGY	1,063	16.1	2,256	18.34
Non-Renewable energy	267	4.0	1,002	8.15
Nuclear Safety	228	3.5	414	3.36
Thermonuclear Fusion	568	8.6	840	6.83
TRANSPORT			240	1.95
TARGETED SOCIOECONOMIC RESEARCH			138	1.12
TOTAL	5,887	89.2	10,686	86.88
2nd Activity				
International Cooperation	126	1.9	5404.39	
3rd Activity				
Dissemination & Evaluation	1% of the budget of each specific programme		330	2.68
4th Activity				
Training & Mobility	587	8.9	744	6.05
Total	6,600	100	12,300	100

Source: SGPN, 1995

Research Personnel. The objective was to strengthen scientific development in Spain, given that some of the less favoured regions of the EU are located here. Evidently, this dissemination must have a favourable repercussion on the socio-economic environment of the country, and this was seen in the negotiations previous to programme approval. Another programme, in which Spain has important participation, is that of Agriculture and Biotechnology, areas of special importance to the Spanish economy. Lastly, stress was laid on the need to adapt part of the Industrial and Material Technology programme to the requirements of the country.

The main objective established by the Spanish authorities regarding the IV Framework Programme is the maintenance of rising economic returns resulting from the different actions: based on a rate of Spanish contribution to European R&D dedicated funds, these can be translated into revenue through future contracts, at least of the same order as that invested. Initially, this evolution is positive given that the results of the III Framework Programme regarding returns were of the value of 5.5%, while in the III, the figure was 6.3%.

Another primordial objective is the establishment of an agile administration structure, facilitating the knowledge of Spanish participation in projects of the Framework Programmes and the economic resources and human capital that may participate in them. This ensures quick response to calls for proposals by means of the formation of business consortia with detailed participation proposals.

Table 1-17 compares the budgets from Framework Programmes III and IV.

To conclude, it can be stated that since the signing of the Maastricht Treaty, the European Commission has tried to attribute greater importance to the clear definition of common R&D policies, raising EU industry to higher levels of competitiveness. An attempt has been made to integrate all science and technology development policies within the Framework Programme and strengthen EU international relations. R&D activities included in the IV Framework Programme have as objective, the strengthening of areas of research related to industry, and in the case of Spain, the identification of what will be high-priority horizontal technologies until the end of the century. This will enable future lines of work to be laid down, increasing the coordination of Spanish S&T policy with that of the other Member States.

Table 1-18: International Programmes with Spanish Participation

Programme/Organisation	Spanish participation (%)	Quota (Millions pts)
EUREKA	23.30	5,200
European Space Agency (ESA)	4.00	15,164
European Organisation for Nuclear Research (CERN)	7.34	4,131
European Synchrotron Radiation Facility (ESRF)	4.00	420
M.V. Laue -Paul Langevin Institute (ILL)	2.10	100
European Molecular Biology Laboratory (EMBL)	7.12	393
European Science Foundation (ESF)	5.82	54
European Molecular Biology Organisation (EMBO)	6.57	75
CYTED (Science and Technology for Development)	—	520
Others	—	47

Source: SGPN, 1994

Table 1-19: International Projects with Spanish Participation - 1994

Project	Other Participants	Budget (Millions of pesetas)	Spanish participation (Millions of pesetas)
CEFIR	E,(NL),F,D,I,UK,UE	31	2
MC2 FLOU	E,(F),I,CH	539	16
ECOIMPACT	E,(S),DK,F	750	82
TAGGER	E,(F),NL,UK,GR	2,600	300
CAROLUS	E,(F)	445	245
OTELSO	E,(SF),UK	650	200
IMALS (DEF)	E,(UK),F,NL	20	4
PACHA	(E),B	550	440
EUROAGRI	(E),I	207	104
RITMICS	E,(F),CH	3,170	1,440
MAINE-TAM	(E),P	395	255
HYBUS	(E),F,UK	504	378
ENZANYM	(E),F	294	168
CAMICO	E,(P)	49	14
MAINE.CONMON	(E),UK,D	246	172
EUROKIWI	(E),P	356	214
EUROENVIRON-CARE	E,(D),F,UK,I,DK,		
VISION 2000	S,SF,A,CH	2,890	30
EUROENVIRON-AQUALERT	(E),P	125	105
EUROENVIRON-MACOM	(E),NL	220	165
EUROENVIRON-SLC PROECESS	(E),IR,DK,CANADA	222	55
ADTT	(E),N	5,000	500
EUROENVIRON-ECOSORB	E,(F)	88	53
EUROAGRI-CELTIFLOR	E,(S),P,UK	200	100
RIDOS	E,(S),P,UK	240	24
LASTECH-HRTD	(E),D	223	133
Total Projects	25	20,015	5,200

Leaders are indicated by parenthesis.

Source: SGPN, 1995

Participation in Other International R&D Programmes

Apart from the Framework Programme of the European Union, Spain also participates in a large number of international programmes and is present in multilateral organisations and centres. The structure of these programmes does not respond to the previously determined lines of common character in the Framework Programmes, but rather satisfies individual necessities of some countries as regards R&D and which are beyond the scope of general European Programmes. Organisational, thematic and administrative structures vary depending on each programme.

International programmes in which Spain participates and the economic contributions made to each one are reflected in Table 1-18. Some of these will be outlined given their importance to the Spanish S&T system.

The first programme that should be highlighted is EUREKA due to its significance for production. Activities are developed mainly through agreed projects, usually integrated into the National Plan for R&D. The programmes of the EUREKA project are usually financed through the entries dedicated to the Industrial Technological Plan of Performance (*PATI*) and through those of the *CDTI*. Spanish participation in these projects is of great importance and in general can be considered highly satisfactory. In 1994, 25 new projects were approved with national participation. 48% of the total of actions are participated in by Spanish researchers and budgetary mobilisation approaches 25% of the total.

Table 1-19, projects with Spanish participation are shown.

Spanish participation in the European Space Agency is also of great interest. ESA activities are centred on the exploration and exploitation of space for civil purposes. In 1993, Spain contributed over 15,164 million pesetas. The *CDTI* is the organisation through which financing is carried out. Given that that the rule of "fair return" is effectively applied, the participation of Spanish industry is favoured. The return accumulated to date by Spain is 102% of its contribution.

Spain is a founder member of the ESA and sixth contributor with a budgetary contribution of 4%. In recent years, there has been gradual normalisation of Spanish relationships with some international organisations. The agreements with the ILL of Grenoble have been re-negotiated, as in the case of EMBL of Heidelberg. Participation has been initiated in the ESRF in Grenoble.

In 1994, negotiations were concluded between Spanish authorities and CERN. By virtue of the agreement reached, the Spanish participation quota was reduced and in exchange, scientific and technological infrastructure was reinforced in the field of Particle Physics. The Centre for High Energy Physics was created and is still operative. The proportion of Spanish personnel has remained steady in the last few years (1.7% of personnel and 9% of scholarship holders), while technological return received by participating companies is 32% (quite a low percentage). It is usual practice that almost all Spanish groups dedicated to High Energy Physics participate in the experiments of the CERN. However, in spite of everything, and given the non-restrictive character of this laboratory at the international level, Spanish presence is only 2.4%.

The last international programme to be highlighted is *CYTED* (*Programa Iberoamericano de Ciencia y Tecnología para el Desarrollo* - Latin American Science & Technology Programme for Development). Given its special characteristics, it is considered avant-garde within the international activities of the *CYCIT*. The importance of this project is not only of a scientific nature, but also because of the close links that Spain maintains with Latin America. The historic links uniting Spanish speaking countries is a well known fact.

CYTED is a programme which encourages cooperation in the field of applied research and technological development to obtain scientific and technological results which may be transferred to the productive systems and social policies of Latin American countries. There are three areas of action: Thematic Networks, Pre-Competitive Research Projects and Innovation Projects.

Spanish participation in this programme is broad, participating in 35 programmes of Thematic Networks and 28 Research Projects, involving a total of 94 companies and public research centres. The total budget for 1994 was 43 million dollars. The number of scientists presently involved in the activities of *CYTED* is 790 out of a total of 7,900 staff. As will be observed, participation is highly significant, keeping in mind that the number of countries involved is 22. Activities developed mobilise resources to the value of 70 million dollars, while the annual budget is 3.5 million dollars.

Table 1-20: Country Participation in actions of the *CYTED* Programmes - 1994

	Thematic Networks	Research Projects	Innovation Projects	Total Activities
Argentina	34	27	7	68
Bolivia	13	3	1	17
Brazil	37	22	2	61
Chile	30	26	3	59
Colombia	29	18	2	49
Costa Rica	23	13	-	26
Cuba	28	12	6	46
Dominican Republic	6	3	1	10
Ecuador	18	8	2	28
El Salvador	10	3	-	13
Guatemala	15	4	1	20
Honduras	11	3	-	14
Mexico	32	25	4	61
Nicaragua	8	3	-	11
Panama	15	5	1	21
Paraguay	13	2		15
Peru	20	12	-	32
Portugal	30	17	1	48
Spain	35	28	34	97
Uruguay	17	11	8	36
Venezuela	35	19	2	56

Source: SGPN, 1995

The results of various programmes have been applied to industry and, thanks to the *IBEROEKA* programme, an initial instrument for the promotion of technological development through cooperation among Latin-American companies has been set up.

Table 1-20 shows country participation in the *CYTED* programme.

As in the case of other projects of an international nature, one of the objectives of *CYTED* is the inter-territorial cohesion of the countries involved in topics related to scientific research. This project has brought about a clear narrowing of north-south collaboration, showing the viability of this and the mutual benefits that may be obtained. The General Assembly of the Programme has guaranteed unique forum for the debate of Latin American R&D policies. In addition, given the diffusion of the programme, cooperation agreements have been signed with organisations such as the Inter-American Development Bank (as observer) and UNESCO.

Spain has also actively collaborated in different committees of the Organisation for Economic Cooperation and Development (OECD), especially in the working group on Scientific and Technological Policy, of which Spain held the presidency from 1991 to 1994. Spain has been involved in other expert committees (Innovation, Patents, Technological Strategies, Productivity and Employment Creation).

Another organisation with which there are effective ties of collaboration regarding scientific and technological development is the United Nations Education, Science and Cultural Organisation (UNESCO). Due to restructuring of the National Spanish Commission for Cooperation with the UNESCO, a *CYCIT* representative was included for the first time in UNESCO decision making bodies in 1995. More than 900 researchers involved in a total of 137 projects have participated in the programmes developed within this framework, with total financing amounting to 847 million pesetas. The contribution of the National Plan to Spanish projects, relative to these programmes, was over 1 billion pesetas for a 3 year period.

In conclusion, Spanish participation in international projects can be considered as satisfactory on the whole, and has brought about a gradual approach of the Spanish S&T system to the levels of competitiveness of the developed countries of the same context. The brief review given of these international programmes shows that there is still a lot to be done. Greater integration in the European Framework Programmes is vital for the improvement of the Spanish S&T system, thus achieving that a greater number of projects are led by Spanish scientists and are therefore better able to adapt to the socio-economic needs of the country. It is also important that companies habitually participating in the resources provided by the R&D National Plan, and integrate their research activities with the objectives of the EU Framework Programme. They should also take advantage of opportunities made available by other international agreements of technological development, in particular with the CERN, ILL and ESRF, given that it is contradictory that great efforts are made for cooperation with these organisations, and this does not translate into contracts with Spanish enterprise.

TECHNOLOGICAL INNOVATION AND DEVELOPMENT IN SPANISH ENTERPRISE

Throughout this presentation of the Spanish S&T system, the main public actions regarding research and development have been outlined. This could lead to the idea that the public sector is the main source of scientific production, and that this source is strictly planned and coordinated. The reality is very different. The main driving-force of the Spanish S&T system is the business sector. It is here where scientific production is stronger and results are more tangible, given the commercial nature of their projects. It should not be forgotten that public action is limited to serving as a driving-force in order to channel scientific research in the most convenient direction for national interests, through the economic development of research areas, which otherwise would not be confronted. It is also the public mission to promote actions which can not be executed by the business sector on an individual basis, due to their complexity and necessary mobilisation of resources. Nor should it be forgotten that the main beneficiaries of scientific production as a whole is enterprise, through the economic returns received. And it is enterprise which takes advantage of technological and innovative improvements to increase competitiveness.

These facts can be deduced from Table 1-21. As can be observed, R&D spending in enterprise is substantially greater than that carried out by public administration.

Table 1-21: Evolution of R&D Spending in the Spanish Business Sector Comparison with other sectors. (1994) (Millions of Pesetas)

Year	Total	Public Administration	Higher Education	Company	Private Non-profit Making Institutions
1990	425,829	90,542	86,721	246,239	2,327
1991	479,372	101,949	106,507	268,434	2,482
1992	539,919	108,035	156,097	272,709	3,078
1993	557,403	111,494	174,342	266,175	5,392
1994	548,154	113,444	173,092	256,316	5,302
1995*	557,910	119,457	176,554	256,316	5,583
1996*	563,490	120,652	178,320	258,879	5,639

forecasts

Source: INE,1997

A description of the evolution of the business S&T system in Spain, its structure and characteristics will now be given as well as data on investment levels and main sectors developing scientific production.

The first relevant fact to be pointed out is that the intensity of the business innovative process is closely related to the level of spending on these activities. According to surveys carried out by the Ministry for Industry, more than 70% of innovative or technological changes take place due to the development of projects within the company and with self financing. Another important percentage is derived from private research collaboration agreements with other companies from the sector. As already pointed out, collaboration with public research centres is not very extensive and only covers from 15% to 20% of this production.

Concentrating on the main reasons leading to research in new technologies, installation of new processes, and development of new products, it can be affirmed that these are fundamentally guided by interest in profitability, and that the investment decisions in R&D are usually considered comparable to those in goods and intangible assets. Due to the criteria of profitability, research within the business framework follows strict planning and organisation criteria in the short run, although not in the mid or long term.

In the business sector, R&D spending is carried out by a small number of companies. According to data of the *INE*, in 1992 alone, 1,753 companies declared that they carried out research activities. Furthermore, only 400 invested quantities greater than 150 million pesetas and monopolised more than 80% of research personnel.

The generation of technology is fundamental in order to raise the level of competitiveness of production, both in the international and national market, and it is an important requirement for the reduction of the imbalance in the scale of payments for acquisition of external technology. In order to achieve these objectives, the National Plan for R&D was established in 1989 and certain fiscal incentives were allowed facilitating research: companies were permitted tax deductions of 30% on capital investment; 15% on spending on intangible assets of research programmes and development of new products or industrial procedures; 30% of intangible assets; and 45% of fixed assets on the increase of spending regarding the previous years.

Spanish companies increased their R&D spending in the period 1982-1992 with an accumulative rate of 20% annually, one of the highest increases in the OECD in that period. The reason for such spectacular growth is the greater liberalisation of the Spanish economy, especially since Spain's joining the European Union. This was a challenge at all levels, forcing national companies into a rapid convergence process with EU partners, in order to avoid being left outside the market. This process of adaptation and convergence was specially significant in SMEs. Before the incorporation of Spain to the European Union, a quite high level of protectionism existed on the part of public administration, a situation with which the SMEs had grown comfortable. Spanish incorporation was followed by a period of adaptation, with gradual tariff reductions being taken advantage of by national companies in order to adapt to new markets. One of the characteristics marking this process of change was the massive acquisition of foreign technology, rather than the drawing up of wide research programmes and own development. This is probably one of the reasons why the dependence on the supply of foreign technology is so high, and the Spanish technology balance, unfavourable. This situation is now changing and the volume of spending on scientific development is continuously increasing. It could be affirmed that Spanish companies are finally up to date and the time has arrived for them to jump on the bandwagon of business innovation and technological development in Europe.

Another business sector of importance is that of public companies. Approximately 20% of the R&D effort is carried out here with 16% of research personnel. Those belonging to the National Institute for Industry (*Instituto Nacional de Industria - INI*) should be highlighted, as well as those of the TENEO group, in the process of being privatised. Such companies are subject to economic guidelines laid down by the government. As regards R&D, Repsol (petroleum company), Telefónica and Renfe (the railway company) are worthy of special mention. These companies which previously operated as monopolies, are now facing up to the liberalisation of their production sectors and rapid privatisation processes. This has led them to implementing all types of innovative changes in order to remain in a market with new competition. As in the case of the SMEs, the initial response to this challenge was the acquisition of foreign technology in order to reach a level of acceptable competitiveness, and to begin developing their own technology, with which they can correct the unfavourable technological balance.

INDICATORS OF THE SPANISH S&T SYSTEM

Science and technology have become essential investments in the economy of any industrialised country. They have been converted into a new sector of activity, pushing and making more dynamic the economic growth of other sectors of the economy. Their influence is comparable to that which sectors such as steel, chemistry, construction, transport, etc. all had in their time.

In recent decades, the experience of industrialised countries has underlined that commercial and productive success falls on technical-economic originality in solving a market need or demand. Originality is innovation resulting from research or effort of technological development.

In Spain, R&D has increased its capacity in the last decade until reaching a stronger position among the group of more industrialised countries.

The recent evolution, at the quantitative level, of the Spanish S&T system will now be dealt with as well as the actual scientific and technological policies, in accordance with the socio-economic context. The basic magnitudes will be analysed on a quantitative and qualitative basis and an effort made to extract as many conclusions as possible.

Such a description of the Spanish S&T system can not be conducted in an isolated manner. The fact that it belongs to the European Union implies, as already indicated on numerous occasions, coordination of the different policies, including those related to scientific research. Therefore, any evaluation made must consider the efforts of the different countries of this background, especially those of the European Union. This is the reason that, whenever possible, comparative analysis will be made of the resources used in such activities, both in terms of time and with other countries.

It must be pointed out that there are great difficulties in obtaining up to date, statistical information on S&T indicators. The real data available to the different organisations involved in scientific research does not usually go beyond the year 1994. That of 1995 and 1996, although concluded, are still only estimates or forecasts and therefore, may be subject to further revision. The most important is not the exact quantification of the parameters defining the system, but rather underlying tendencies; growth, effort by public administration in strengthening certain sectors, business interest in certain fields, etc.

R&D Expenditure and Financing: General Indicators

The first parameters to be analysed are the financial resources used in R&D. The first difficulty arising in these cases is the definition of the statistical concept of R&D. Due to this, public administrations have made a great effort in homogenising this concept in such a way that comparative analysis may be carried out, both in time as with other countries, regarding resources used for such activities.

The following table shows the evolution of expenditure on R&D and Gross Domestic Product (GDP) in Spain.

Table 1-22: Expenditure on R&D in the period 1987-1996

Expenditure on R&D	1987	1992	1994	1995*	1996*	Annual accumulative rate (%)
Current pesetas (Billions of pesetas)	231	540	548	557	563	12.9
Constant pesetas 1986=100 (Billion pesetas)	110	186	174	187	188	6.8
Purchasing Power ($ Millions)	2,263	4,589	4,455	4,641	4,691	10.2
GDP growth. Current pesetas						8
GDP growth. Constant pesetas						3
R&D Spending over GDP (%)	0.64	0.92	0.85			

* forecasts
1994 data are forecasts of the INE and of the SGPN

Source: INE (1997), OECD (1994), SGPN, MINER

As can be observed, the rate of real growth in expenditure on R&D is measured in constant pesetas, reaching a value of 6.8 percentage points in 1994, equivalent to 12.9% in current pesetas. On comparing this with growth in GDP, it can be observed that the increase in ERD (Expenditure on Research and Development), doubles the increase of this last factor (in current pesetas). This implies that Spanish society in general, is making an important effort to strengthen the scientific sector of the country, even above its normal level of growth. It can therefore be deduced that special awareness exists regarding the processes of technological innovation and their importance in the socio-economic context.

Another indicator of interest is to be found in the distribution of spending carried out and in the origin of the funds dedicated to R&D activities in Spain.

Table 1-23: Spending Carried Out in R&D Activities in Spain - Breakdown by sectors. (Millions of pesetas).

Year	Total	Public Administration	Higher Education	Company	Private non-profit making institutions
1990	425,829	90,542	86,721	246,239	2,327
1991	479,372	101,949	106,507	268,434	2,482
1992	539,919	108,035	156,097	272,709	3,078
1993	557,403	111,494	174,342	266,175	5,392
1994	548,154	113,444	173,092	256,316	5,302
1995*	557,910	119,457	176,554	256,316	5,583
1996*	563,490	120,652	178,320	258,879	5,639

* forecasts

Source: INE, 1997

Table 1-24: Origin of R&D Funding in Spain (Billions of Pesetas)

	1986	1987	1988	1989	1990	1991	1992	1994
Foreign	3	3	7	16	19	22	27	34
Central Administration	70	88	107	120	150	167	181	287
Regional Administration	15	18	21	25	35	45	55	-
Universities	11	12	13	14	15	18	20	-
Sector Company	97	108	138	162	186	221	255	220
Private, non-profit making institutions	2	2	2	2	2	2	2	5
Total	198	231	288	339	407	475	540	548

Sources: Central public administration: General Budgets of the State.
Autonomous administration and Universities: Estimates of the SGPN
Business sector: INE.
Foreign: Estimates of the INE and of the SGPN

The capital available for the development of all types of projects related to research and development comes fundamentally from the business sector and public administration, these also being the main users of funds.

Probably the best way to appreciate at a glance the scientific research effort being carried out, is to capture in graphic form the trends in ERD growth in relation to the net growth of the economy in terms of Gross Domestic Product.

Figure 1-16: Breakdown of the recent evolution of R&D spending

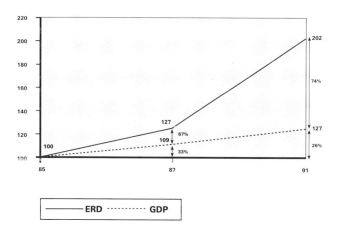

The objective of this figure is to show, whether GDP or ERD (in terms of percentage) experiences greater growth. 1985 will be taken as base year with a value of 100 for both GDP and ERD. In 1987, ERD acquired a value of 127, while GDP reached a value of only 109. In 1991, the difference is even greater, ERD being 202 while GDP was 127. Analysis of the figure shows that the growth in investment effort in scientific areas can be divided in two; that which is inherent to the growth of the economy, that is to say, the greater GDP, ERD will increase in equal measure (at least in theory) if there is not anything promoting it; and that derived from greater promotion of investment in scientific research activities. In 1987, ERD growth was due in 33% to growth in the GDP and in 67% to greater investment. The situation was even better in 1991, where only 26% is attributable to GDP and 74% to greater effort. The drive by Spanish society in these activities goes further beyond pure economic growth, which is higher than most OECD countries. This obviously causes rapid convergence with the countries of this background, raising Spain's position in the EU.

This increase in effort may appear optimistic but there is still a lot to be done. The real indicator that measures the effort made by society regarding scientific research is the percentage of GDP dedicated to R&D. For this reason, and in spite of experiencing growth, Spain is clearly still in a position of inferiority regarding the developed countries of the same context. In Table 1-25, this data is shown in greater detail.

Table 1-25: Expenditure on R&D over GDP (%)

Country	1989	1990	1991	1992	1993	1994
Germany	2.88	2.81	2.83	2.81	2.43	2.33
France	2.34	2.40	2.42	2.45	2.45	2.38
EU (Average)	2	2.01	2.1	2.3	1.95	1.91
Spain	0.75	0.81	0.87	0.92	0.97	0.84
Italy	1.24	1.35	1.35	1.4	1.26	1.19

Source: INE, OECD

In spite of the fact that the differences with other countries, especially with the more developed ones, are decreasing, the ERD in relation to GDP is still greater than Spain; 0.84 in Spain in comparison with 2.33 in Germany or 1.91 the EU average. This difference may appear very substantial, but as already indicated the growth differential of the rate of investment in R&D, grows at a higher rate in Spain than in most other OECD countries, in such a way that in the not too distant future, efforts carried out in scientific research may be compared. This fact may be corroborated by comparing the average annual accumulative rates of ERD.

Table 1-26: Average accumulative rate of ERD (%)

Country	1985-1987	1987-1990	1990-1993
Germany	7.5	7.1	7.1
France	5.4	9.1	10
EU (Average)	-	7.8	8.2	9.1
Spain	15.4	16.6	16.2
Italy	8.7	9.9	10.1

Source OECD, INE

The average, accumulative, annual rates are higher in Spain in comparison to the other countries. Comparison with countries of long standing traditions in research, such as Japan, shows a wide margin of difference. This series of data clearly reflects the convergence effort being carried out.

Table 1-27 outlines the evolution of R&D expenditure according to sector and funding. The first part reflects the percentage dedicated to R&D expenditure by the different agents involved in the process (public administration, higher education, enterprise and private, non-profit making institutions). Likewise, the financing of R&D activities is shown in percentage in the same sectors including foreign investment. Both tables are accompanied by the accumulative average growth rate.

Table 1-27: R&D Expenditure. Growth by sector

Expenditure on R&D	Average Annual Accumulative Rates (%)					
	1984	1987	1992	1984-87	1987-92	1984-92
Public Administration	26.0	25.2	21.5	20.7	15.1	17.5
Higher Education	22.2	19.0	18.9	16.2	19.6	18.1
Companies & Private Non-Profit Making Institutions	51.8	55.8	59.6	25.7	21.7	23.3
Total	100.0	100.0	100.0	22.4	19.7	20.9
Financing of R&D Activities (%)						
Administration	50.0	51.1	48.5	23.3	18.2	20.3
Central Administration	44.4	38.1	35.2	16.3	17.4	16.9
Regional Administrations	3.2	7.8	9.5	65.1	25.7	41.3
Universities	2.4	5.2	3.8	58.7	10.7	29.2
Companies & Private Non-Profit Making Institutions	49.2	47.6	46.9	21.1	19.3	20.1
Foreign	0.8	1.3	4.6	44.2	64.5	55.5
Total	100.0	100.0	100.0	22.4	19.7	20.9
Transfers of the public sector & abroad to the business sector						
R&D Expenditure (%)	2.6	8.2	12.7	85.0	33.3	53.4

Source: INE and SGPN

Taking financing first, it is necessary to highlight the high growth rates of the regional governments, due to the important political and economic transfers granted to them in areas related to science. Although funds dedicated by central administration have diminished, they have done so with the help of an important increase in financing by EU funds, which is also a public body.

Spending in current pesetas carried out by the different agents increased from 90 billion pesetas in 1982 to 533 billion pesetas in 1993. As shown in Table 1-27, the most significant growth rate corresponds to the business sector, reaching values greater than 20%. That of the other sectors as a whole is slightly above 16%. In this same table, it can also be appreciated that approximately 57% of spending was carried out in 1992 by companies in comparison with 49% by the same sector in 1982. It can therefore be concluded that the effort made by such agents is increasing. This should initially be considered normal, due to two fundamental reasons. The first is that normally, companies are the end receiver of the innovation processes and, secondly, are those that must most quickly converge with foreign competition.

The increase in investment rates in the business sector is not unsusceptible to underlying public help. In recent years, the effort of public administration in increasing R&D grants has been considerable, and since Spain's joining the

European Union and through the R&D Framework Programmes and ERDF, companies have been able to benefit from new sources of financing. The annual growth of such transfers reached a growth rate of 33.3% in the period 1987-1992. This supposes a change in the level of transfers from 2.6% of total R&D spending in 1986 to a level of 12.7% in 1992.

The setting up of such aid to enterprise, both in the case of public administration and in that of international organisations, has permitted the adaptation of the structure of the origin of funding and its application to that of EU countries in a short period of time. An example of this is shown in Table 1-28. The percentage of transfer to the Spanish company is around 12% of ERD. Other countries such as Germany, the United Kingdom or France register similar percentages. In any event, modifications in the spending structure tend toward rapid international homogenisation, of particular importance since 1987 and 1988.

Table 1-28: Structure of the Origin & Application of Funds in Other Economies. (%)

Expenditure on R&D	Germany	France	Italy	United Kingdom	United States	Japan
Public Administration	13.2	23.3	24.9	14.5	12.4	7.5
Higher Education	14.4	14.3	18.0	15.4	15.6	17.6
Companies & Private Non-Profit Making Institutions	72.4	62.4	57.1	70.3	72.0	75.0
Financing of R&D Activities						
Public Administration	34.1	48.1	51.3	36.9	48.2	18.0
Companies	63.3	43.9	44.5	50.0	49.5	73.1
Foreign	2.1	7.4	4.2	9.8	—	0.1
Other Sources	0.5	0.6	—	3.2	2.3	8.8
Transfer of the public sector and from abroad to the business sector						
R&D Spending (%)	8.6	17.9	12.6	16.9	20.2	-7.0

Source: OECD

Production

The evolution of the main indicators in the business sector, as well as their relation to the S&T system of the European Union will now be described and is reflected in Table 1-29. As previously indicated, R&D spending in the business sector has grown considerably, moving from 2.3% of the EU total in 1987 to 3.3% in 1992. Looked at from the point of view of percentage on GDP, the increase is also substantial going from 0.35% in 1987 to 0.46% in 1992.

Regarding human resources, a similar evaluation can be made. In 1987, personnel dedicated to R&D represented 2.6% of the EU total, increasing to 3.7% in 1992. This increase is also associated to the fact that, within enterprise there

Table 1-29: R&D Activity in the business sector.

	1987	1992	Growth (%)
N° of companies carrying out R&D activities	1,140.00	1,753.0	9.0
Expenditure on R&D in the business sector (Billions of Pesetas)	126.00	273	16.7
Constant pesetas (Billions of pesetas 1986=100)	108.00	168	9.2
% of EU($ Current purchasing power)	2.00	3.3	-
% of Spanish GDP	0.35	0.4	-
Enterprise financing of national spending (Billions of current pesetas)	108.00	235.0	16.8
R&D Personnel in sector	20,361.00	28,590.0	7.0
% of EU	2.00	3.7	-
Researchers in enterprise	6,835.00	11,593.0	11.1
% of R&D Personnel	33.00	40.5	-
Patents			
N° of patents applied for in Spain	23,391.00	48,900.0	15.9
By residents	1,741.00	2,101.0	3.8
By foreigners	21,649.00	46,799.0	16.7
N° of Patents applied for abroad	2,263.00	6,886.0	24.9
Dependency rate (application for patents foreign / residents)	12	22.3	-
EU Dependency rate	2	4.0	-

Source: SGPN, 1995

is an increasingly greater quantity of personnel incorporated into technology development activities. At present, the annual growth rate is 7%. However, the percentage of researchers with medium or high level qualifications of total R&D personnel is still inferior to the EU average. This indicates that the level of science in Spanish companies has still not converged with the average level of the European Union.

As for scientific production in the business sector, the number of patents requested increased from 23,391 in 1987 to 48,900 in 1992; the figure almost doubled in 5 years. The dependency rate, defined as the relationship between patents requested by non-resident researchers and those requested by researchers resident in Spain, shows a significantly growing tendency, much higher than that of the EU as a whole. Doubtless, the reason for this is the increase in the number of patents requested by non-residents due to Spain's incorporation into the European market.

Patents requested abroad increased from 2,263 in 1987 to 6,886 in 1992. Of these, 284 were requested in the European Patent Office. Although this is significant, it should not be forgotten that the figure represents only 3% of total patents requested in the EU. Initially, it may appear that the balance of scientific production is poor, but if compared with real scientific production, according to data of the Science Citation Index (SCI) shown in Tables 1-30 and 1-31, it can be observed that the value is 6.1% regarding the EU total. The conclusion is that there is no real concern for the legal protection of research results in Spain. At present, the responsible public bodies are stressing the importance of the protection of such results and of the consequences for the innovation process.

Table 1-30: Results from the Spanish S&T System.

Results	1987	1992	Rate of Growth
Scientific production*	8,816.00	13,860.00	9.50
Quota of production regarding world total	1.22	1.83	-
Coverage Ratio of Technological Balance	0.18	0.25	-
Patents requested abroad	2,263.00	6,886.00	24.90
Patents requested in the European Patent Office	149.00	284.00	13.80

Source: OECD, SCI, MINER, SGPN
*Scientific production measured by the SCI index

Table 1-31: Results of the Spanish S&T system. International comparison.

Results	Spain	EU	OECD
Scientific production*	13,860.00	225,251.00	638,519.00
Quota of scientific production - Spain	—	6.10	2.20
Patents requested abroad	6,886.00	22,749.00	1,131,841.00
Quota of foreign patents - Spain	—	3.00	0.60
Index of self-sufficiency (Patent application.Resident/National.)	0.04	0.20	0.37

Source: OECD, SCI, MINER, SGPN
*Scientific production measured by the SCI index

As already indicated, R&D spending in the business sector basically concentrates on a reduced number of companies, and therefore, on a reduced number of industrial branches. The following table gives a sectorial breakdown of R&D spending in Spanish companies in 1991, in accordance with the OECD statistics available. As can be deduced from the data obtained, more than 70% of total expenditure is carried out in the sector of manufacturing industries, in particular; metal, electricity and chemicals. In recent years, almost 30% of total R&D spending has been concentrated on the sectors of office material, the pharmaceutical and the aerospace industry, attributing special importance to these sectors. The spectacular growth of non-manufacturing branches must also be pointed out; likewise the services sector, in which spending in 1991 was greater than 10% of the business sector total. Nevertheless, it should be pointed out that this branch is closely linked to industrial activity through research associations.

Table 1-32: Sectorial Comparison: Spain - EU

	1987		1991	
	Spain	EU	Spain	EU
Aerospace industry				
Percentage of R&D total expenditure	8.7	10.4	6.6	11.3
Exports/Imports	0.71	1.21	0.38	1.05
Electronic and Electric industry				
Percentage of R&D total expenditure	14.4	26.9	17.4	23.3
Exports/Imports	0.41	0.94	0.40	0.8
Office Material and Computers				
Percentage of R&D total expenditure	7.1	4.4	5.9	4.3
Exports/Imports	0.38	0.47	0.36	0.39
Pharmaceutical Industry				
Percentage of the R&D total expenditure	8.8	7.8	8.8	8.8
Exports/Imports	0.91	2.07	0.65	1.70
Other Manufacturing Industries				
Percentage of R&D total expenditure	42.2	41.8	39.8	43.8
Exports/Imports	0.87	1.31	0.76	1.10

Source: OECD

It may also be deduced from this table, that the spending structure of sectors in Spanish enterprise, coincides in great measure with that of the European Union.

Another interesting aspect is company size and the R&D effort of each. As shown in Table 1-33, large companies with more than 1,000 workers represent less than 10% of the total of Spanish companies, but carry out 40% of R&D spending. Evolution from 1987 to 1994 shows increasingly more decisive participation by SMEs in research projects, in detriment to the concentration experienced in the larger companies. At present, in companies with less than 250 employees (73% of the total), 31% of business spending is centred on R&D. This indicates that SMEs, which in Spain possess a weight of 64% (in mobilised resources), are becoming aware of the need to set up specific R&D departments in order to develop their own innovative processes or new products. In spite of everything, large companies continue to be the main driving-force as far as technology development is concerned.

Table 1-33: Evolution of the business effort in research - %

	1987	1994
Nº of Companies		
< 250 workers	66.1	73.5
250 to 500	15.1	13.1
500 to 1000	8.0	6.4
1000 or greater	10.8	7.0
Total	100.0	100.0
R&D Expenditure (%)		
< 250 workers	25.6	31.9
250 to 500	12.3	12.1
500 to 1000	10.7	16.8
1000 or more	51.4	38.6
Total	100.0	100.0
Researchers (%)		
< 250 workers	36.5	32.4
250 to 500	12.6	13.8
500 to 1000	10.1	15.0
1000 or more	40.8	38.8
Total	100.0	100.0

Source: INE, 1997

With regard to the origin of capital, the group of companies with foreign capital executes 40% of the processes of business R&D. This contrasts with the fact that, in number, they only represent 19% of the total. In turn, they employ 32% of researchers and 37% of the total of personnel with R&D related functions. Companies with entirely national capital carry out the remaining 40% of scientific research (measured in financing) and in spite of lesser effort, show significantly higher growth rates. The pace of growth is 10% in comparison to the 6% of foreign capital. This series of data is shown in the following table.

Table 1-34: Evolution of the indicators of R&D of the business sector according to the percentage of foreign capital in the company.

Company type	1989		1992		Annual Accumulative Growth
	N°	%	N°	%	
N° of companies	1,341	100.0	1,753	100.0	9.3
Without foreign capital	939	70.1	1,254	71.6	10.1
Less than 20% foreign capital	51	3.8	68	3.9	10.1
20% to 50% foreign capital.	73	5.4	97	5.5	9.4
More than 50% foreign capital	278	20.7	334	19.0	6.3
R&D Expenditure (Millions of pesetas)	191,153	100.0	272,709	100.0	12.6
Without foreign capital.	87,934	46.0	119,367	43.7	10.7
Less than 20% foreign capital.	25,826	13.5	31,011	11.4	6.2
20% to 50% foreign capital.	6,360	3.3	13,574	5.0	28.7
More than 50% foreign capital.	71,033	37.2	108,757	39.9	15.2
Researchers	9,394	100.0	11,539	100.0	7.2
Without foreign capital	5,115	54.4	6,355	54.8	7.5
Less than 20% of foreign capital.	802	8.5	972	8.4	6.6
20% to 50% foreign capital.	307	3.3	517	4.4	19.0
More than 50% foreign capital.	3,170	33.7	3,750	32.3	5.7
R&D Personnel	25,865	100.0	28,590	100.0	3.4
No foreign capital	12,903	49.9	14,194	49.6	3.2
Less 20% foreign capital.	2,352	9.1	2,667	9.3	4.2
20% to 50% foreign capital.	923	3.6	1.309	4.6	12.3
More than 50% foreign capital.	9,687	37.4	10,420	36.5	2.5

Source: INE and SGPN

A description of the Spanish business sector would not be complete without mentioning the cooperation links between companies and public research centres. This is perhaps the weakest point in the data indicated. Contribution by

Table 1-35: R&D in Public Administration & Higher Education.

	1987	1992	Growth
Spain			
R&D Expenditure			
Current pesetas (Millions of pesetas)	101,855	264,132	20.9
Constant pesetas			
(Millions of pesetas 1986=100)	96,307	180,251	14.3
Current dollars (millions)	1,010	2,244	17.3
R&D Personnel	27,900	44,231	9.6
Researchers	19,625	29,827	8.7
European union			
R&D Expenditure (Current $.)	27,076	39,011	7.5
R&D Personnel	519	614	3.4
Researchers	257	282	1.9

Source: SGPN, 1995

Table 1-36: R&D in Public Administration.
Spain.

	1987	1992	Growth
Expenditure on R&D			
Current pesetas (Million pesetas)	58,188	108,035	13.2
Constant pesetas			
(Million pesetas, 1986=100)	55,019	73,726	6.0
Current dollars (Millions)	577	918	9.7
ERD as a % of public sector total	57	40	
R&D Personnel			
Personnel with Full Dedication	12,507	16,678	5.9
R&D personnel as % of public sector total	44.8	37.7	
Researchers			
Researchers with Full Dedication	4,525	7,660	11.1
Researchers as a % of public sector total	23	25	
Resources			
R&D Spending per researcher			
(thousands of $)	127	120	-1.2

Source: SGPN, 1995

companies to the R&D carried out in public research centres alone is 4% of total research expenditure. Only in the area of university-industry links, do the indicators improve somewhat. According to 1992 data, collaboration rose to 7%, with an annual growth rate of 75% in terms of total spending. In the period 1987-1992, total spending of collaboration between enterprise and higher education centres increased from 1,100 million pesetas to 11,500 million, an almost insignificant figure if compared with total R&D spending. Various conclusions may be drawn from this data. The first is the lack of knowledge or trust of enterprise in Spanish universities. The feeling exists that both branches of research go their own way, when the most logical thing would be close collaboration. It can be concluded that there is little industrial application of the scientific research carried out in higher education. In spite of everything, and as already indicated, the growth of these links is significant, mainly because a large number of universities have begun to avail themselves of modern and sophisticated means, and are starting to manage enterprise resources.

There is still much to be achieved, especially due to the low capacity of Spanish companies to work with public research centres. This is the greatest handicap in Spanish business S&T in comparison with European competitors.

SCIENTIFIC ENVIRONMENT

The "scientific environment" refers essentially to a group of public research centres and universities. Results are frequently measured by the number of publications. In order that they have a certain value, only those appearing in international databases will be taken into consideration. This quantification is carried out by the Science Citation Index. According to their figures, from 1987 to 1992, Spanish production of scientific articles has increased with an accumulative rate of approximately 10%. Improvement in the quality of magazines must also be considered, in addition to the increase in number, and the number of references made to such publications.

Tables 1-35 and 1-36 show the evolution of R&D spending in public research centres and universities for 1987-1992. Data on R&D personnel, researchers with full dedication, is also given and comparison is made with EU figures.

Four fundamental conclusions can be drawn from these tables. The first is that growth of the investment level in the public sector is greater than in the business sector. Secondly, within R&D public spending, higher education is daily acquiring greater importance in relation to that of public administration. Thirdly, R&D spending in the public sector is 50% of national spending and lastly, growth experienced by Spain both in public spending on scientific research and in personnel dedicated to R&D activities is the highest within the European Union, at a clear distance to the rest of the member states.

This data may seem optimistic, but one of the most serious problems of the Spanish S&T system is the gap between the scientific environment and that of production. Initially, this is not numerically quantifiable. Relations between the

Table 1-37: R&D in Higher Education. Spain

	1987	1992	Growth
R&D Expenditure			
Current pesetas (Millions of pesetas)	43,667	156,097	29.0
Constant pesetas (Millions of pesetas, 1986=100)	41,288	106,525	20.9
Current dollars (millions)	433	1,326	25.1
ERD as a % of public sector total	43	60	—
R&D Personnel			
Personnel with Full Dedication	15,393	27,553	12.3
R&D personnel as % of public sector total	55	62	
Researchers			
Researchers with Full Dedication	15,100	22,167	8.0
Researchers as a % of public sector total	77	74	
Resources			
ERD per researcher (thousands of $)	28,7	60	15.8

Source: SGPN, 1995

Table 1-38: Structure of R&D Funding in Industry (%)

	Year	Public Administration	Companies	Foreign	Other sources
European Union					
	1987	16.6	77.4	5.8	0.1
	1992	12.3	79.9	7.6	0.2
Spain					
	1987	13.8	83.6	2.4	0.2
	1992	11.4	80.5	8.0	0.1

Source: OECD, INE, SGPN

scientific environment and other international research organisations are good, but their work is mainly stimulated by the publication of scientific articles and not by cooperation with enterprise. Motivation due to the technological aspects of scientific results is not sufficient.

Indicators of the Financial Environment

The importance of the financial context is not limited exclusively to the facet of financing of different research projects. The same process of resource assignment is also a key element in the global policies determining an S&T system; it is greater or lesser difficulty in obtaining resources, which sectors experience greater ease when obtaining financing, in what conditions it is received, etc.

Table 1-38 shows the structure of R&D funding within Spanish enterprise. The majority of financing originates from the company's own funds. Only 22% is external financing. In 1992, the weight of financing of various public administrations reached 11.4%. The general trend is that the weight of these interventions is increasingly smaller, but that the orientation greatly defines the global R&D policy of national interest. Another fundamental aspect reflected in the table is the existence of parallelism between Spain and the European Union concerning the origin of R&D financing in the business sector. As previously indicated in this chapter, Spain has experienced gradual convergence in the structure of the origin of funds and their application in the S&T system.

Another fundamental aspect of the public financing of R&D, is the preferred type of company. Public administration (Table 1-39) attempts to strengthen research in smaller companies (less than 500 workers), this being the sector which dedicates less resources to R&D. SMEs experience great difficulty in getting appropriate financing for R&D from venture capital societies. Such societies, which negotiated funds to the value of 160 billion for R&D activities in 1992, prefer to aid companies in expansion phase, rather than newly created ones. Preference is also for larger companies over small or medium sized ones. Funds assigned were therefore carried out in a very low risk manner by these societies, guaranteeing their economic interests rather than possible social benefit.

Public administration also directs investment toward public companies rather than private ones, and those which have entirely Spanish capital rather than those with foreign capital.

In addition to direct actions by public administration for the development of R&D activities, (subsidies, credit, etc.) there are also indirect ones of no lesser importance. These are fiscal incentives, presently serving as a driving-force in the Spanish socio-economic fabric as far as scientific research is concerned.

Table 1-39: Percentage of Companies and R&D Activities

	ERD	Public Financing	Relative public Financing (%)
Size			
<100	16.5	23.6	1.4
100 to 500	23.3	30.5	1.3
>500	60.2	45.9	0.8
Total	100.0	100.0	
Capital			
National	43.8	67.3	1.5
Foreign	56.2	32.7	0.6
Total	100.0	100.0	
Company type			
Public	18.4	35.4	1.9
Private	81.6	64.6	0.8
Total	100.0	100.0	

Source: SGPN, INE, 1992

Regional Indicators

The autonomous nature of regions within the Spanish state, as established in the Constitution, has notably conditioned the S&T system. Within the area of political responsibility corresponding to the regional governments, each has initiated actions in the area of R&D promotion on an individual basis, creating the corresponding bodies and endowing them with the appropriate legal framework to enable them to carry out their activities. In recent years, practically all regional governments have set up active policies for the promotion of R&D. Basically, this was a response to the need felt by regional administration to support technological development through specific initiatives in this area. Therefore, their budgets in the last decade reflect a high increase in R&D funding.

The first consequence of investment decentralisation is its somewhat heterogeneous distribution, creating a series of imbalances and insufficient inter-territorial cohesion in this type of activities. In Spain, this provoked the appearance of two types of regional government; wealthier governments which dedicate a high level of funds to R&D activities, promoting important enterprise development, and poorer ones which depend in greater measure on central administration activities. Fortunately, this unbalanced situation is gradually improving.

Table 1-40 shows R&D expenditure by regional governments for the period 1987-1992. At the beginning of the period (1987), Madrid made up half of national spending on R&D, followed at a distance by Catalonia, the Basque Country and Andalusia. These four regions together made up more than 34% of the total number of personnel employed in this type of activity. R&D expenditure by central public administration is also territorially concentrated, taking into account that 62% goes to Madrid., followed by, again at a distance, Andalusia and Catalonia with 8% and 10% respectively. This indicates that decentralisation of research in institutes of public administration is still an outstanding task, especially in the less favoured regions, where the State only contributes with 23% of total national spending.

Spending by the university faculties, technical schools and technological institutes appears to be better distributed, although it still shows high concentration. Madrid and Catalonia each have 16%, followed by: Andalusia, 14%; Valencia, 12%; Castile Leon, 6%; and the Basque Country, 5%. It should be highlighted that the total of the remaining 12 less favoured regions, only receives 52% of national spending.

Also significant are the regions of the Balearic Islands, the Canary Islands and Extremadura, where research by enterprise is very low. This is due to the fact that almost all research is developed in the public sector. Contrary to this situation is that of Catalonia, the Basque Country and Madrid. The structure of their spending is comparable to that of other regions in industrialised countries. The majority of enterprise with innovative bent is concentrated in these areas, as is the highest concentration of public research centres, especially in metropolitan areas.

At present, part of public R&D spending is decentralised and the regional governments are in charge of its distribution. In fact, the regional governments, being in a better position to know what the particular needs of their own S&T systems are, have promoted successful R&D actions with increasing effectiveness. In 1993, budgetary headings controlled by the regional governments represented more than 18% of total public funds dedicated to technological development and innovation. In global terms, this means that since 1988 spending realised by these public organisations has doubled.

The same line of heterogeneity in the research effort being examined is also to be found in regional government financing. In the following figure, regional government financed R&D is shown in comparison with the gross value added of its production. Catalonia possesses the best index with 0.17%, while other regions such as the Balearic Islands, the Canary Islands or Cantabria present almost valueless percentages. The case of Madrid is atypical. Initially, it would appear that the ratio of regional government financing / Gross Value Added at Factor Cost is too low in comparison with the other regional governments, considering that its total level of R&D investment is the highest in Spain. The reason for this is that most of the capital investment in this region is at the cost of the central government and enterprise, as it is here that the highest number of research centres are to be found. The ratio given above is not significantly high, although financing by regional governments is quite considerable.

Table 1-40: Territorial Distribution of ERD - 1992

	Total Spending		Public Administration		Universities	
	Millions of Pesetas	%	Millions of Pesetas	%	Millions of Pesetas	%
Andalusia	45,088	8.2	9,357	8.2	24,768	14.3
Aragon	13,514	2.5	3,248	2.9	5,263	3.0
Asturias	8,011	2.5	1,747	1.5	3,952	2.3
Balearic Islands	1,943	0.4	390	0.3	1,379	0.8
Basque Country	42,634	7.8	992	0.9	7,297	4.2
Canary Islands	13,356	2.4	3,414	3.0	9,635	5.6
Cantabria	4,729	0.9	1,016	0.9	2,549	1.5
Castile and Leon	25,877	4.7	1,644	1.4	13,241	7.7
Castile-La Mancha	4,684	0.9	891	0.8	1,381	0.8
Catalonia	109,748	20.0	12,466	11.0	30,009	17.3
Extremadura	4,690	0.9	1,349	1.2	3,089	1.8
Galicia	14,069	2.6	2,569	2.3	7,191	4.2
Murcia	7,357	1.3	1,877	1.7	3,696	2.1
Madrid	203,250	37.1	68,589	60.5	29,862	17.3
Navarre	7,788	1.4	293	0.3	3,057	1.8
Rioja	1,339	0.2	270	0.2	315	0.2
Valencia	34,642	6.3	3,362	2.9	21,740	12.6
No Specific Region	5,428	1.0	0	0.0	4,658	2.7
Total	548,153	100.0	113,443	100.0	173,091	100.0

Table 1-41: Percentage Distribution of Total ERD by Enterprise (1994)

Region	%
Andalusia	4.3
Aragon	2.0
Asturias	0.9
Balearic Islands	0.1
Basque Country	12.6
Canary Islands	0.1
Cantabria	0.5
Castile and León	4.3
Castile-La Mancha	0.9
Catalonia	25.8
Extremadura	0.1
Galicia	1.7
Madrid	40.6
Murcia	0.7
Navarre	1.7
Rioja	0.3
Valencia	3.6
Total Spain	100.0

Source: INE, 1997

In Table 1.40 spending by private, non-profit making institutions (3,078 millions) is not included in total expenses. Their activities are carried out mainly in Madrid and Catalonia.

Figure 1-17: Regional Government R&D Financing /Gross Value Added at Factor Cost. 1992

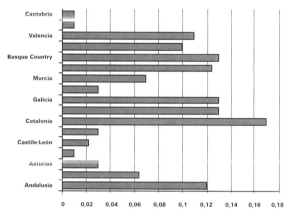

Source: SGPN

In conclusion, there is high imbalance among Spanish regional governments regarding technological innovation, although the situation has tended towards improving in the last 10 years. Both national and EU programmes are presently carrying out important tasks of cohesion with the objective of driving scientific research in lesser developed regions. According to recent statistics, in the past 5 years the level of inter-regional cohesion has increased 25% and is now one of the most important rates in the European Union.

There are several reasons for this convergence. First to be highlighted is the improved application of both national and EU funds and the cohesion task carried out. Second is the rapid spread of university and public research centres in areas where they did not previously exist. Last is the application of the Structural Funds, the primary objective of which is the economic convergence of less favoured regions.

PERSPECTIVES OF THE SPANISH S&T SYSTEM

The fundamental conclusion to be drawn from the evolution of the Spanish S&T system, is that today it is much closer to those of its EU partners than a decade ago. The situation is now one of change. The second National Plan for R&D and the III Framework Programme of the European Union are concluded, fulfilling their objectives for better or worse. Perhaps now would be a good time to question whether or not maintaining the same structure and operating system is appropriate to achieving the final objective of convergence with the other EU countries, as well as tackling future challenges. In brief, it is fundamental to reflect on how relations between enterprise and public administrations should evolve regarding the mobilising effect of the latter.

The restructuring supposed by a new national plan should not exclusively lead to the re-channelling of resources dedicated to certain research areas to other new ones. Rather, there should be development of new actions without losing the previously obtained results. The most preferable evolution must move toward re-orientation of existing actions. This chapter proposes to outline the re-structuring that, according to the institutions responsible for the development of scientific policy in Spain, must be undergone by the S&T system. Lastly, the economic backgrounds which could make scientific convergence possible will also be outlined.

Ideal Orientation of the Spanish S&T System

In the section dealing with the indicators of the Spanish S&T system, it was indicated that one of its defects was insufficient collaboration between enterprise and the public sector; not only on an economic basis but also at the level of scientific production. The first fundamental step to be taken is the development of such co-operation actions, ensuring that the areas of action of both R&D systems evolve jointly and not separately, as has been the situation to date. The other aspect which must evolve is the financial environment. At present, it is not easy to get R&D funding, unless the solvency of the applicant is unquestionable: it is not easy to persuade financial entities to get involved in research activities with the level of risk being dealt with here.

The evolution of the S&T system will depend on the ease with which its actions can mobilise economic means and the capacity to adapt to the new relations among them.

In order to channel relations among the basic R&D actors, the *OTRI/OTT* network was set up with the objective of disseminating the results of different areas of research in areas where they could be of use. This network is still not the efficient interface initially sought after between enterprise and the public research centres, although activity has increased. At present, it is resulting in more effective links within the scientific environment and not within enterprise. It is therefore necessary to create more efficient communication links between both sectors, and within these, between technology and production. A result-orientated concentration in the projects is also necessary: an efficient channel of information would be useless, if the information provided did not have suitable application. The solution to this problem could perhaps come about by the creation of communication nodes at the level of central administration or, in particular, regional administration, or even inside the different sectors of enterprise. Such nodes should be able to communicate in a permanent, agile and effective way, and the access by enterprise, should be clearly useful.

As for the problem of funding R&D activities, it is necessary to make capital risk societies aware of the need for their involvement in the development of scientific research. Again it is essential that the different levels of public administrations drive the creation of entities which can provide funds to technologically innovative companies.

These objectives are those of the III National Plan for R&D; evolution toward a more favourable situation rather than rupture with all that has been previously carried out. The following step will be to describe how the global objectives of the Plan (promotion, planning and coordination of scientific research) will be modelled.

Promotion of Scientific Research

The increase in public and private resources made available to the S&T system should lead to an increase in physical and human capital, thus assuring its survival, and in particular, future results. Spain is carrying out an important degree of effort as regards convergence in R&D funding. The objective is to reach EU average investment rates in coming years. At present, such rates are approximately 2% of GDP. To do so, it will be necessary to maintain the trend initiated in 1986: greater growth (percentage wise) in ERD than in GDP. Following the current rhythm of growth, the existing gap between the EU average and Spain would be eliminated in the year 2015 approximately. However, if this objective is set for within the next 3 years, it will be necessary for the percentage of the GDP dedicated to R&D activities to reach values around 1.3%. The European Union urges its Member States to increase its global investment in scientific research. It is proposed that the distances separating the EU as a whole from its most direct competitors, Japan and the United States, be reduced. If the effort is considerable at the economic level, it must be even more so with respect to the level of personnel involved in R&D activities, as the distance separating Spain from the other countries is even greater.

This increased R&D effort must be carried out in a compensated manner. As shown in the previous section, the rates of growth of investment levels were much more important in the sector of public administration than enterprise. This in itself constitutes an imbalance, greater affecting Spain than the other European countries. It is therefore the task of the III National Plan for R&D to redress this situation as much as possible.

As for the convergence in level of personnel involved in R&D activities, measured through the "R&D Personnel/Active Population" indicator, the distance with the rest of the countries is still quite considerable. In spite of the fact that the distances are decreasing gradually, this is not being carried out in the most appropriate way. It is an established fact that in time, the Spanish S&T system continues responding to the models of evolution of the lesser technologically developed countries of the Union; structures more intensive in labour than in infrastructure tend to be adopted. Such expansion has resulted in Spain moving away from the general trajectory in recent years, contrary to what happened in Italy for example.

If, as already mentioned, the percentage of the GDP dedicated to R&D should be approximately 1.3% for the year 1999, then the rate of growth in research personnel should be located at around 6%, with the objective of reaching 8% of the active population in the same year. Thus the demand for research staff taking into account the increase in population will be covered.

Evidently, desirable growth must affect the Spanish S&T system, both globally and on an individual basis, regarding each of the agents involved. It does not only consist of adapting the level of investment and number of research personnel in line with the GDP of the country, but that such growth take place in a harmonic and planned way, without the disequilibrium shown here.

Planning of Scientific Research

The objective of R&D planning was to bring the areas of scientific research closer to the socio-economic needs of the country. It is therefore fundamental that the third phase of the National Plan for R&D includes within its general lines, the specific contributions that can be made by the agents involved. Obviously, basic research or long term projects, which could initially appear to be less applicable but which are indispensable in the long run, should not be forgotten. However, it is important to emphasise projects of more immediate application in the desire to bring scientific research closer to production.

During the first two phases of the National Plan, the high-priority objective was the application of funds to strategic sectors of special national interest through the national programmes. This model of available resource distribution is very similar to that employed by the EU in assigning funds under the Framework Programmes. The main problem is that large differences exist between both concepts (National Plan-Framework Programme). Firstly, the Framework Programme of the EU represents a small percentage of total R&D expenditure within Europe, and secondly, this model is only of use when the objective is to promote in general a weakly developed S&T system. Although some of the deficiencies of research in Spain have been pointed out here, it should not be forgotten that initially it is a consolidated system in expansion. Therefore, the objectives of the National Plan must be otherwise; it could be said that such objectives must correct the imbalances that inevitably appear in all socio-economic aspects experiencing very rapid growth. Firstly, a general increase of R&D effort in its two aspects, spending and researchers, is required with the objective of reaching real convergence with the EU average. Secondly, to favour the regional cohesion of scientific research which at present in Spain is showing high disequilibrium. Thirdly, to favour relations between research developed in enterprise and that carried out in public research centres, in such a way that both grow in parallel and not following different paths. Fourthly, that the dissemination of the scientific and innovative results is efficient, in such a way that its integration and application in the national socio-economic context are carried out in a quick and agile manner. Lastly, that R&D results overall play an appropriate role as a basis for the different sectorial policies.

The number of programmes to be carried out in the third phase of the National Plan has been significantly reduced. Initially, the objective sought is not a concentration of efforts in certain areas, but an appropriate dosage of available resources to achieve the goals laid out above.

The areas of basic research should not be abandoned, nor should the real needs of enterprise as regards R&D. Scientific-technical knowledge, demanded by all innovative processes, can only be adequately transferred by an efficient approach to the necessities of the production sector. Such an approach will allow room for basic, quality research and areas of a strategic nature, emphasising the dissemination of knowledge, the transfer of results and their future application to emerging innovative processes. A scheme of this type, designed to contribute flexibility in its development but rigidity in its objectives, will permit better coordination with the policies aimed at enterprise and implemented by the regional governments; the latter is important given that activities close to the market are being dealt with.

This flexibility, together with a clear determination in the objectives to be achieved, must be captured in a series of conclusions from an operative point of view. They can be summarised in three aspects. Areas of research linked to production objectives should be developed through the National Programmes, considered more appropriate for the agile dissemination of R&D results in the short and mid-term. Each high-priority strategic objective will be of a different level of importance according to the impact on the production sector or the development of the research group carrying it out. Emphasis on the different areas of actions is no longer determined by the importance of these, but also by the capacity of who will develop them. Lastly, and perhaps this is one of the most fundamental aspects that the National Plan for R&D must show regarding flexibility, is the adjustment of the effort dedicated to each area of research in response to the degree of receptivity demonstrated by the corresponding sector of production.

R&D Coordination

The third aspect of organisations in charge of the development of the National Plan is the coordination of the actions of the agents responsible for research and development. This coordination was carried out at three levels: coordination of national actions, territorial coordination and international coordination.

The need for new scientific and technical knowledge to solve pressing problems (desertification, hydraulic resources, rehabilitation of the disabled, industrial innovation, etc.) can be seen at present in various sectors, granting growing importance to the coordination of scientific policy with sectorial coordination.

As seen in the section dedicated to the coordination of the Spanish S&T system, coordination actions were initiated in addition to those related to the Sectorial Plans which constitute the main part of the actions. Such actions were developed on a thematic basis as well as horizontally and vertically, and acquired special importance given that they included for the first time, companies presently carrying out the largest number of research projects. To date these actions can be considered as incipient and their main mission consists in acquiring the experience necessary to develop future plans of greater scope. It is necessary to highlight coordination of the National Plan with *PATI* of the *MINER* by means of various mechanisms, especially through projects in collaboration with enterprise, integrated projects with the participation of a large number of actors and areas of research, and special actions within the framework of EU R&D Programmes.

Actions with reference to coordination through the sectorial programmes, have been limited to the Ministry of Education and Science, to the Institute for Agricultural and Food Research and Technology and the Fund for Health Research. To date, the objectives of the National Plan as a whole have not been considered other than coordination actions at the sectorial level.

This may be the area where a policy of more effective coordination is more necessary with the regional. The General Council for Science and Technology has not been able to implement a systematic and effective policy in this field. Although the experience gained through Regional Government Programmes in the National Plan has been extensive, it is still insufficient to expand it to all possible environments. The appropriate design and implementation of mechanisms of co-administration and co-financing has been achieved, but the complex action mechanism has not allowed them to function as a basic coordination instrument. This is not so in the case of the European Structural Funds, already shown to an effective instrument of cooperation between state and regional government, as well as an excellent means of increasing territorial cohesion, and complementing the National Plan in all its aspects.

The III National Plan for R&D must gradually correct this series of deficiencies regarding coordination. Its actions will be oriented in line with the following:

To strengthen the instrumental character of the Plan as scientific, technological and administrative support of government sectorial policy, by means of coordination with the actions of Ministry Departments.

To take into consideration sectorial policies designing the objectives of the National Programmes outlined by the CICYT

To expand and improve actions of thematic coordination that allow flexible action and provision of quick responses to the scientific, technological and socio-economic needs in general of the country.

To intensify and rationalise coordinating policies with regard to large centres on the advice of the Advisory Committee of Large Scientific Centres.

To design and set up coordination mechanisms with the governments of the autonomous regions, which are flexible, agile and easily administered and which can be adapted with ease to the needs of each region.

To increase coordination between pluri-regional programmes and regional programmes of the European Structural Funds, and integrate these actions into the infrastructure programmes outlined by the National Plan or by regional plans.

Source: CICYT

Activities of coordination at sectorial, autonomous regions and external level foreseen in the third phase of the National Plan for R&D will now be dealt with.

Sectorial Coordination

Coordination of sectorial policies acquires greater importance every day as social demands in certain areas (environment, health, etc.) increase, and programmes begin to show an important supranational component, in particular with the rest of the European Union. The initiation of new areas of coordination action, based on joint planning, administration and financing, with ministry departments is essential, as is the improvement of existing channels of communication.

Ministry of Health and Consumption

Criteria for evaluation and administration have been established through a single call for all the programmes related to this Ministry; National Health Programme; Sectorial Programme of the Fund for Health Research; and the Sectoral Programme of General Promotion of Knowledge (in topics related to health sciences).

Ministry of Public Works, Transport and Environment

Spain's deficit in areas of research related to environment is significant in comparison with its EU competitors. The problem of the shortage of hydraulic resources demands the development of new areas of research to ensure correct distribution and use. Actions have therefore been developed in the following matters:

> *Setting up of the National R&D Programme on Climate, as technological and scientific support of the National Programme on Climate, agreed and negotiated with the Secretariat of State for Environment and Housing and, in particular, with the National Institute of Meteorology.*
>
> *Support to the National Hydrological Plan, through the setting up of a National Programme on Hydric Resources. This Programme will be agreed with the Secretary of State for Territorial Policy and Public Works, the Secretary of State for Environment and Housing and other units of the MOPTMA, MAP and MINER, competent in this area.*

Source: CICYT

Ministry of Agriculture, Fisheries and Food

Coordination with the National Institute for Agricultural and Food R&D has to date being carried out effectively. The high-priority objective is to maintain the areas of action developed so far. These are :

> *Gradual deepening of coordination with the INIA.*
>
> *Setting up of the National Programme for Marine Science and Technology, in collaboration with the Spanish Institute of Oceanography.*
>
> *High-priority attention for short term sectorial needs of the agricultural industry through actions co-ordinated with the General Secretary for Food. The integrated project on olive oil will be put into effect in an immediate phase.*
>
> *Consideration of specific actions in areas such as forest fires and forestry policy in general, erosion and dessertification, and others of special environmental interest, all that related to the National Programme for Environment and other Sectorial Programmes of possible development with the MOPTA.*

Source: CICYT.

Ministry of Social Affairs

The specific functions of the programmes of this Ministry is to attend to the needs of society which are not covered by the production sector: necessities of less favoured communities (old age, disabled, socially isolated, etc.) The following actions have been planned:

> *Initiation of a specific mobilising action (integrated project) related to the application of new technologies to the problems of the disabled, agreed with the National Institute for Social Services (Instituto Nacional de Servicios Sociales - INSERSO)*
>
> *Integration into the National Plan, such as Sectorial Programme, of the research activities carried out by the Institute for Woman, through the Annual Plans for Equal Opportunities for Women (action begun in 1995).*

Source: CICYT

Ministry of Culture

Coordination actions with this Ministry are guided fundamentally by the search for technical solutions to preserve the artistic heritage of the country. These actions demand very wide coordination since a great part of responsibility in this area has been transferred to the regional governments.

Ministry of Defence

The nature of the programmes carried out by the Ministry of Defence demands the coordination of numerous researchers and resources. The mechanism habitually used is that of integrated projects. The viability of continuing and expanding the *MINISAT* Project in coordination with the National Aerospace Research Programme is being studied, the agreement of which will be carried out through the National Institute of Aerospace Technique (*INTA*). The work carried out by the oceanographic ship, Hesperides and the Spanish base in Antarctica will also continue.

Ministry of Education and Science

Main coordination actions have been promoted from the Secretariat of State of Universities and Research. Given the vital nature of the activities of this Ministry to the S&T system, the development of coordination of its activities to date has been effective, especially regarding the Sectorial Programme of General Promotion of Knowledge. In the future, it is hoped to go into more depth in the following areas:

> *To continue the Sectorial Programmes of General Promotion of Knowledge and of Training and Improvement of Research Personnel, revised in accordance with the general guidelines of the III National Plan for R&D.*
>
> *To continue, on behalf of the DGICYT, administration of the National Programme for Training of Research Personnel, revised in accordance with the employment perspectives for researchers in the public and private sector, the specific needs of the industrial sector and the correct use of programmes of the European Social Fund.*
>
> *To coordinate R&D actions related to sport and promoted by the Higher Council for Sport with the National Plan.*
>
> *To initiate specific actions on educational technologies.*

Source: CICYT

Ministry of Foreign Affairs

In recent times, the growing internationalisation of scientific research has obliged a large part of the coordination of research activities to be carried out through the Ministry for Foreign Affairs. In general it can be concluded that it is necessary to tighten relations with the institutions responsible for the development of scientific activities (EU, OECD etc..). In particular, it is necessary to make relations more agile between the *CYCIT* and the Ministry regarding High Energy Physics in order to take advantage of Spanish participation in the CERN. Close cooperation is also necessary in actions related to the *CYTED* programme.

International Coordination

As already mentioned, a large part of the coordination of the transnational component of the Spanish S&T system is carried out through the Ministry for Foreign Affairs. The mission of the *CICYT* in this field is to assure a high level of subsidiary, and that the expected financial returns are received through appropriate coordination. Another of the most important objectives is to assure a high number of projects lead by Spanish research teams, and to increase Spanish participation in those already being carried out.

As is to be expected, the Framework Programme of the European Union will monopolise the main part of the coordination effort. The activity of administration can be considered adequate to date, but greater integration of the National and Regional Government Programmes with objectives common to the Framework Programme, is necessary. As is normal, both types of areas of activity are less independent and therefore greater coordination between them is necessary.

The development of the IV Framework Programme of the European Union is making improved coordination essential among the different levels of Spanish public administration, with the objective of unifying approaches and possessing coherent postures when negotiating national participation in these scientific projects. In such a way, it ensures that the large research areas of the country are reflected within the objectives of the European Programme.

Spanish participation in the EUREKA and *CYTED* programmes must be highlighted as must its participation in CERN and the ESA, given that to date results are satisfactory. However, the fundamental objective continues being an increase in participation in all.

Regional Governments

As has already been seen, the General Council for Science and Technology was unable to carry out the coordination foreseen of the development and technology innovation policies of the regional governments. In order to correct this problem, commissions and work groups are being organised to draw up concrete proposals in this area. The objective is to achieve appropriate coordination of infrastructure actions and personnel training.

The channelling of relations between central and regional government foresees that the following is carried out:

To establish quick and operative systems for information exchange, with the IRIS Network as a fundamental element.

To include objectives of special interest for regional governments among the priorities of National Programmes that will contribute to their financing in smaller or greater quantities according to each case.

To coordinate National Plan calls with homologous calls of regional governments; this may imply anything from the simple coordination of information to the launching of joint calls in certain actions.

Regional government participation in integrated projects approved by the CICYT, and collaboration in project administration and financing.

Setting up integrated projects on the initiative of one or several regional governments, which would suppose co-financing higher than in the previous case by the region, and maintaining the Plan's general principles of action (open and public calls, possibility of participation of all Spanish groups, evaluation, etc.). In this case administration of the project would be shared by the regional government and the CICYT.

A variant of this mechanism is the co-financing of integrated projects of action in a certain industrial sector, including a first phase of technological audit, with the participation of MINER and the CDTI

Joint participation in the administration (and implementation if necessary) of large scientific centres, in accordance with the criteria approved by the Permanent Commission of the CICYT.

Source: CICYT

Enterprise

With the purpose of profiling R&D coordination in enterprise, and in particular with public research centres, inefficiently developed to date, two new mechanisms have been developed in the III National Plan. The first is systematisation in the grouping of programmes by objective, and in second place, the creation of the figure of Enterprise Promoter/Observer (EPO).

As for the first of these mechanisms, it is understood that the majority of programmes related to enterprise can be categorised into two main groups, the first of which is "thematic." In this category all areas of research considered to be of high priority are included, but which are susceptible to immediate, practical application. The second category is "finalistic", in which actions are dedicated to the development of scientific projects which have as objective, fulfilling existing demand in such a way that they unite scientific nature with the greatest possible value added. The latter will be granted greater priority than the former, and both public administration and enterprise will be jointly involved in their management and financing.

This mutual collaboration can be carried out through the different execution mechanisms described in previous sections: Integrated Projects, should various companies, scientific groups, and public research centres be involved; Agreed Projects if participation by enterprise is only a part of its development; a *PETRI* action if participation by the company is limited to the transfer of technology from a public research centre. Lastly, in the III National Plan a new figure or mechanism, the EPO, is included which complements the three outlined above.

The EPO is an entity (company or unit of public administration) from which it is proposed that resources be devoted to the finalistic objective of the National Plan and which commits itself to participating in the development of the corresponding project during its execution. Participation can have different degrees of intensity and implication, from mere observation and follow up to active collaboration. As this figure is new, it is still not possible to evaluate its effectiveness.

In the III National Plan for R&D a new programme titled Promotion of the Design and Implementation of the Science-Technology-Industry System (*Fomento de la Articulación del Sistema de Ciencia-Tecnología-Industria - PACTI*) is developed and serves as a basic tool for coordination and collaboration between public research centres and the production sector or enterprise. The main objective is adapted to a great extent to OECD recommendations, which lead to the promotion of the structure of science, technological and production environments and promote effective orientation and use of scientific and technological knowledge and capacity by the production sector and society in general.

Adequate development of links between the scientific and technological environments and production and the assurance of efficient transfer of results is necessary in order to achieve such objectives. The best way of doing so is to take into consideration both of these aspects from the very moment in which the project is conceived. Such a spirit is clearly implemented in the two basic mechanisms of collaboration between public administration and enterprise: *PETRI* actions and Agreed Projects.

The results of the *PETRI* actions have been highly satisfactory to date, although it is fundamental to assure the productive partner's co-financing in the entirety of the project, and to guarantee the participation of the implied technology sectors to assure the transfers of results to SMEs. It is proposed that a greater number of projects from the production sector be involved in the National Plan with the clear objective of this leading to *PETRI* actions.

It is foreseen that the same areas of action will be continued regarding Agreed Projects, while improving links with projects which are more pre-competitive in nature and also with those related to industrial development (*PATI*, EUREKA, *IBEROEKA*, ESA). Within the principles that inspire Agreed Projects another element to be introduced is association with the programmes for the incorporation of researchers to enterprise. Such objectives must necessarily be accompanied by an improved financing system of the projects to be carried out. In other words, try to improve the financial conditions of reimbursable aid, giving priority to companies in Objective 1 regions (the least developed) among other factors.

As for follow up mechanisms of both actions, *PETRI* and Integrated Projects, which the III National Plan seeks to establish, these will be directed at a richer dialogue between enterprise and public R&D agents.

Financial Background

As already indicated, the level of Spanish spending on R&D, grew in the period 1987-1994 to a average accumulative rate of around 13%, thus reducing the difference with other countries of this context. This supposes in figures an increase from 425,8 to 542,3 billion pesetas or 27,3% in current pesetas.

Part of this spending has been administrated by the National Fund, which in spite of its small volume still makes possible detailed control of application according to areas of action and scientific-technical objectives. Their overall budget and distribution of funds are subject to rigorous control by the *CICYT*, its Advisory and General Council, the Council of Ministers and the Senate Parliament Commission.

Basically this is an R&D development mechanism, and the fact that it is administrated by the *CICYT* makes it a perfect instrument for the coordination and planning of various public administrations. This coordination effect has materialised in 51% of Spanish researchers being currently involved in activities related to high-priority areas of the National Plan, and it has been capable of mobilising the triple of external resources toward the same activities.

With reference to the future, it is considered evident that the Spanish S&T system must continue growing in order to converge with the select club of the European Union. However, this growth need not be uniform in all contexts, but rather must be centred on company growth. The financial scenario foreseen for 1999 is shown in Table 1-42, and illustrates that R&D financing by enterprise should reach 53% of the total spending, a similar percentage to that of EU average in 1992.

Table 1-42: Perspectives of the evolution of R&D expenditure

Year	ERD	GDP	Enterprise Funding		Public Funding		Central Administration		National Fund
	%	%	Billions Pesetas	%	Billions Pesetas	%	Billions Pesetas	%	Billions Pesetas
1994	0.84	45	244.0	49	265.7	70	185.9	4.0	21.9
1995	0.85	46	271.0	48	282.8	69	194.5	3.6	21.5
1996	0.88	47	308.1	47	308.1	68	209.5	3.8	24.9
1997	0.92	48	352.5	46	337.8	67	226.3	4.0	29.3
1998	0.96	50	405.7	44	357.0	66	235.6	4.2	34.0
1999	1.00	53	465.7	41	360.2	65	234.1	4.4	38.6

(1) On public financing
(2) On total R&D spending

Source: SGPN, 1995

In general these hypotheses suppose a consolidation of the economic and industrial recovery in the mid-term, and what is more important, gradual consciousness of the Spanish entrepreneur on matters regarding research and development. The new orientation of the third phase of the National Plan should contribute to this, reducing as much as possible the difference that still separates Spain from the EU average.

NOTES

1 Prologue to Plan Nacional de Investigación Científica y Desarrollo Tecnológico. Published by *Comisión Interministerial de Ciencia y Tecnología*, 1988

REFERENCES

Boletín Oficial del Estado (1986) Ley de Fomento y Coordinación General de la Investigación Científica y Técnica.

Comisión Interministerial de Ciencia y Tecnología (1993), Cooperación Tecnológica Industrial. Cuaderno 3.

Comisión Interministerial de Ciencia y Tecnología (1995), Sistemas Regionales de Innovación. Cuaderno 5.

Comisión Interministerial de Ciencia y Tecnología, El III Plan Nacional de I+D. 1996-1999.

Comisión Interministerial de Ciencia y Tecnología. Los Programas de I+D de la Comunidad Europea.

Comisión Interministerial de Ciencia y Tecnología. Memoria de Actividades del Plan Nacional de I+D en 1993.

Comisión Interministerial de Ciencia y Tecnología. Memoria de Actividades del Plan Nacional de I+D en 1992.

Comisión Interministerial de Ciencia y Tecnología. Resumen de la Memoria de actividades del Plan Nacional de I+D en el Cuatrienio 1988-1991 y Perspectivas Futuras.

Consejo Económico y Social(1994), Situación y Perspectivas de la Industria Española: Sesión del Pleno

Dorado R. et al (1991), Ciencia, Tecnología e Industria en España. (Madrid: Fundesco)

European Commission (1992) De l'Acte Unique à l'après Maastricht. Les Moyens de nos Ambitions. Documento COM (92) 2000 final. (Brussels)

European Commission (1992), Comparison of Scientific and Technological Policies of Community Member States: Spain.

European Commission (1993), SMEs and Community Actions in the Area of R&D: COM 93. Brussels,

Eurostat, Europe in Figures (Fourth Edition).

Fundación 1º de Mayo (1990), Ciencia y cambio Tecnológico en España.

IMPI (1993) La Pequeña y la Mediana Empresa. Informe Anual.

International Monetary Fund (1996), Perspectives of the World Economy.

Journals, Articles and Pamphlets published by the Secretariat of the National Plan and the CDTI.

La Politique Communitaire de la Recherche et du Developpement Technologique audelà de 1992. Brussels. 1991

OECD (1992), Technology and the Economy: The Key Relationships.

OECD (1994). Science and Technology Policy: Review and Outlook.

OECD (1995), Fiscal Measures to Promote R&D and Innovation: Trends and Issues. (Paris: OECD).

OECD (1995), Statistiques de Base de la Science et de la Technologie.

OECD (1996), Economic studies of the OECD: Spain

Office of Official Publications of the European Communities (1995), General report on the Activity of the European Union.

Oficina de Transferencia Tecnológica, La Red OTRI / OTT.

Oro, Luis A. I and Sebastián Jesús (Ed), Los Sistemas de Ciencia y Tecnología en Iberoamérica, Colección Impactos.

Oro, Luis A., El sistema español de ciencia y tecnología. Colección Impactos

Research and Technological Developement. Achieving Coordination through Cooperation. Brussels 1994

Sanchez P. and Folguera A. .El Sistema Financiero Español como Impulsor de la Innovación: Un análisis Comparado con los Países de la OCDE.

The Economist Intelligence Unit, Country Profile: Spain. 1995-1996.

Web Pages of the main Spanish public research centres.

PORTUGAL

Chapter 2

PORTUGAL

INTRODUCTION

The recent evolution of the Portuguese S&T system has been parallel to that of Spain: the two countries joined the European Union at the same time, facing similar challenges. Since the date of its adhesion, Portugal has been involved in a more and more competitive market, where the security resulting from restrictions or taxation of imports has gradually disappeared. Portuguese society, through its public institutions, recognised the need to create a new S&T system in order to provide the country with a sufficient level of competence in order to converge with the EU average.

The consequences of that convergence, as far as scientific research is concerned, have been diverse, and came about as a result of the new economic order. Sectors, (especially primary ones) which in the past were fundamental in the national economy were pushed aside to make room for others (services sector) with a high capacity of penetration, both in the national market and in the European Common Market.

Such developments brought about a new series of necessities that had to be resolved by the national S&T system. As laid out in the Portuguese Constitution:

> *Scientific research must be promoted and protected by the State (Article 73.4).... Both in the economic and social spheres the role of the State should be... to develop scientific and technological policies, preferably in those areas related to the development of the country, in order to eliminate foreign dependence. (Article 81).*

The Portuguese State has taken on a fundamental role in this aspect. As will be seen, the impetus of public institutions regarding research is high, in contrast to a production sector somewhat less sensitive in these matters. The former is due to the financial injection of the European Union aimed at developing the system. Portugal has known how to take advantage of this new source of funding, and some of its more important sectors have been able to come up to the level of their European neighbours.

A fundamental aspect of any S&T system are the corresponding public institutions involved in its development, consolidated and situated within the socio-economic framework of the country. In the case of Portugal, this has not yet been fulfilled given that the State is presently making special effort and creating new organisations at the ministry level, responsible for the development, planning and coordination of the new system. These new organisations are recent and it is still too early to judge whether or not they fulfil the expectations created in the scientific community. The system has also been provided with a new legal framework, through which the participation of the main economic agents in the elaboration of programmes and high priority research areas, has been strengthened.

The convergence effort being made by Portugal is bringing about positive results. In spite of being the second lowest EU country in terms of R&D spending over GDP, it is also one of the quickest growing. It is sufficient to point out that at the end of the 80's, Portugal experienced the highest public investment growth in research in the OECD. This effort is corroborated by its increasing participation in the R&D Framework Programme of the European Union, a level which is above that which it should be, taking into account the size and development of its S&T system.

Evidently, the challenges presented by the system are similar to those shared by all of the countries in this context, especially the lesser developed technologically. The following should be highlighted:

- R&D processes intensive in human capital and not in infrastructure or financing.
- Excessive weight of public participation in research compared with that of enterprise.
- Dependence on EU funding, in contrast to internal public spending that must be increased.
- A ratio of scientist/population lower than that in more developed countries
- Lack of projects with wide reaching impact on the European economy, due to lower assignment of resources.
- Centralisation or scientific research around large urban nuclei, leaving aside less favoured regions.

In spite of everything, the prospects of growth are quite optimistic. Although the institutions have not been long in operation, a global consciousness of the need for R&D in order to be competitive exists in Portugal. This is the fundamental starting point toward lessening the dependence on foreign technology and becoming a new driving-force for scientific research.

LEGAL AND INSTITUTIONAL FRAMEWORK OF S&T POLICY IN PORTUGAL

Legal Framework

Law 144/96 of Scientific and Technological Development constitutes the legal framework of the S&T system in Portugal. Here the principles of system reform and the organisations that must carry it out are laid down. New mechanisms for action, constituted by diverse organs, are determined and the correct participation and coordination among them is guaranteed.

The reasons for this law include that of the necessity to develop an S&T system which serves as a driving-force for the socio-economic environment of the country as well as a guarantor of modernisation and of the economic, social and cultural development. Fundamental actions are defined and the creation of national plans for specific R&D are laid out in order to achieve these basic objectives and develop appropriate policies. In concrete, the law includes:

- Definition of general high-priority objectives of scientific policy: the development of knowledge; the development of scientific research and the transfer of derived results to the production sector; the improvement of researcher training, both at initial level and throughout the development of their activities (continuous training); the strengthening of the scientific and technological community; the development of relations among the different entities responsible for R&D, both public and private; the dissemination of a scientific culture, etc.
- Framework for the development of long term R&D policies is laid down (ten year periods with partial revisions every three years). Evaluation and follow up of these policies is carried out annually and is the responsibility of the parliament.
- Definition of general guidelines for the planning of scientific research.
- Common approaches for the evaluation of research are defined.
- Regionalisation of the S&T system with the objective of decentralising tasks, so that implementation and development of precise actions are carried out by the regions themselves, with appropriate national planning and coordination.

In order to facilitate the development, planning, coordination and administration of scientific activity, the Law presses the Government to carry out the reforms necessary to restructure the competent R&D organisations.

Institutional Structure

The Law of Scientific and Technological Development grants the Government the capacity to establish the necessary, institutional mechanisms for the development of the S&T system in Portugal. The set of institutions and relations among them are reflected in Figure 2.1 below.

Ministry for Science and Technology

This Ministry, recently created (1995), is the body with primary responsibility for the elaboration, planning, coordination and follow up of the Portuguese S&T system. It is presided over by the Minister of Science and Technology, while other Ministries related to areas of R&D are also represented. Its main mission is the definition of the scientific policies that must be implanted in the country

Figure 2-1: The Portuguese S&T System

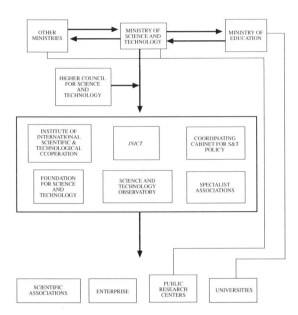

and the establishment of appropriate mechanisms for their development. It coordinates Portuguese participation in international, bilateral or multilateral projects, both within and beyond the EU, incorporating these projects into the structure of the national programmes and assuring appropriate scientific returns.

The Ministry is also responsible for the assignment of resources to public, R&D programmes through the appropriate mechanisms, as well as for directing the activities of some of the more prestigious public research centres, among which should be highlighted: Institute for Scientific Tropical Research (*Instituto de Investigação Científica Tropical - IICT*); Technological and Nuclear Institute (*Instituto Tecnológico e Nuclear - ITN*); Scientific and Cultural Centre of Macao (*Centro Científico e Cultural de Macau - CCCM*); Academy of Sciences of Lisbon (*ACL - Academia das Ciências de Lisboa*). It also negotiates specified private funds dedicated to the programmes integrating the national programmes.

Another mission is one of guiding the policy of researcher training at all levels, proposing measures for the promotion of employment and facilitating researcher mobility in both research and production. In order to carry out its functions, the Ministry of Science and Technology collaborates closely with: the Higher Council for Science and Technology (*CSCT-Conselho Superior da Ciência e Tecnologia*); Co-ordinating Cabinet for Scientific and Technological Policy (*Gabinete Coordinador da Política Científica e Tecnológica*); Science and Technology Observatory (*Observatório das Ciências e das Tecnologias - OCT*); Institute of International Scientific and Technological Cooperation (*Instituto de Cooperação Científica e Tecnológica Internacional - ICCTI*);

Science and Technology Foundation (*Fundação para a Ciência e a Tecnologia - FCT*). This set of institutions and others of special importance in the implementation of the Portuguese public S&T system are outlined below.

Higher Council for Science and Technology

The Higher Council for Science and Technology (*Conselho Superior da Ciência e Tecnologia - CSCT*) was created in 1986 in order to substitute the National Council for Scientific and Technological Research, originally the organisation responsible for the main advisory tasks to the government on scientific research, but which was unable to develop its activities efficiently. The newly formed Council took over responsibility for such assessment, carrying out its activities in a more acceptable manner. Since 1996, due to the restructuring of the Portuguese S&T system, such consultancy has undergone deep restructuring, adopting new functions which will be outlined below. (Decree-Law nº 145/96, 26 August 1996).

The *CSCT* works in coordination with the Ministry of Science and Technology. It is a collegiate body whose main mission is to advise on S&T related topics at all levels in Portugal, concentrating in particular on the following: coordination; financing; planning; evaluation; international relationships; and policies to be developed in general. Under its own initiative, it formulates opinions and proposals based on its own studies, with the objective of improving scientific research in Portugal. The Council expresses opinions, views and recommendations on questions relating to scientific and technological policy, either as a result of its own initiative or proposals from other entities. Such questions are:

- Basis of scientific and technology policy.
- Pluri-annual planning of research policies.
- Public budgets for the development of science and technology.
- European scientific and technological policies in coordination with national policies.
- International scientific and technological cooperation.

The President of the Council is chosen for a two year period by the representative members of the institution. To effectively develop their work of assessment, the *CSCT* has eleven members among its personnel who are government elected from among prestigious personalities in the area of science and technology, and the majority of whom have previously held high ranking positions in public administration in areas related to research. It also has representatives from: the Coordinating Council for Higher Polytechnical Institutes (*Conselho Coordenador dos Institutos Superiores Politécnicos - CCISP*); the Council of University Rectors (*Conselho de Reitores das Universidades Portuguesas - CRUP*); and various non-profit making, science related institutions. As representation by enterprise is also fundamental for the appropriate development of its activities, the *CSCT* has six delegates representing companies which are active in R&D. Social agents are represented through employer associations, trade unions and political parties holding seats in the *Assembleia da República*.

In accordance with its President, the *CSCT* participates in high level committees, where concrete aspects of scientific research in Portugal are analysed, or in evaluation committees.

National Council for Scientific and Technological Research

The National Council for Scientific and Technological Research (*Junta Nacional de Investigação Científica e Tecnológica - JNICT*) is, along with the Ministry of Science and Technology, the main Portuguese institution in matters related to scientific research. Its creation dates from 1967 (Law 47/791) and since then it has been responsible for the coordination and promotion of scientific and technological research, in accordance with the guidelines laid down by the Ministry of Science and Technology. Coming under the control of the Minister, it is closely linked to the various ministries responsible for R&D: Education, Culture, Defence and Economy. To date, the work carried out has been of great importance to the Portuguese S&T system. However, its functions will be greatly reduced by the *Leis Organicas das novas instituçoes* to be published sometime in mid 1997, which will bring about the transfer of its responsibilities to the Ministry for Science and Technology.

Among the functions which were originally granted to the *JNICT* by the Law for Scientific and Technological Development, the following may be highlighted:

- Realisation of studies to define policies necessary for S&T development.
- Proposal of general guidelines for public financing of scientific research, with the objective of increasing the productivity of public research organisations.
- Preparation of national R&D plans in perfect coordination with development plans already being carried out. These plans will identify general guidelines that should be adopted long term, as well as short term, concrete actions.
- Preparation and up-dating of resources available which are dedicated to scientific research.
- Promotion and funding of researcher training at all levels.
- Development of scientific programmes.
- Collaboration with the organisations responsible for national defence on scientific aspects of interest to the armed forces.
- Coordination with the Ministry of Foreign Affairs for the participation of Portugal in international, scientific research projects.
- Promotion and participation in the structures and networks responsible for the dissemination of scientific results.

Due to the long period of time in which it has been developing its activities, the *JNICT* has undergone successive reforms in its structures and ways of action. One of the most important, apart from that expected in 1997, was that of July 1994, which was an attempt to ensure its flexibility and the more active participation by the scientific community in its activities and also tried to guarantee rigour in procedures and transparency in the decision making processes. Decision making organs were thus restructured as follows:

- Board of Directors.
- General Council.
- *INVOTAN* Commission.
- Specialised Commissions.
- Inspection Commission.

The General Council is the organisation within the *JNICT* which aids the Board of Directors in relation to its functions of planning, coordination and promotion of R&D activities in Portugal. The Council consists mainly of representatives from the Ministries competent in scientific research, representatives of public economic administration (at departmental level), and various of the most outstanding figures of scientific research in Portugal.

Specialised Committees were set up to assist particular scientific areas on a more individual basis, as well as the planning, coordination and development needs of scientific research. Such Committees consist of a member of the General Council, one or more of the national delegates responsible for coordination with EU programmes related to *JNICT* activities, and prestigious specialists, representing research institutions. Their mission is to cover, from a technological point of view, different sectors of economic and scientific activities: environment, biology, biotechnology, agricultural sciences, engineering, human sciences, social sciences, marine sciences and technology, earth sciences, defence, energy, physics, mathematics, materials, chemistry and chemical engineering, health, information technologies and telecommunications, production technology, automation and robotics.

The main functions of these Specialised Commissions are as follows:

- To advise the *JNICT* Board of Directors on the lines of activity to be followed for scientific and technological development in each area of specialisation.
- To propose specific initiatives to the Board of Directors to be implemented in the area of specialisation.
- To propose methods and approaches for the validation of proposals of R&D projects, job data bases and other S&T activities.
- To announce verdicts on candidatures to R&D projects.
- To follow-up studies of projects developed within the *JNICT*.

The members of these Specialist Commissions are named on a three year basis.

The *INVOTAN* Commission, set up within the General Council is responsible for the coordination of cooperation in the area of science and technology within NATO, through the *INVOTAN* programme, which will be referred to in greater depth at a later stage. The *JNICT* Board of Directors, the Ministry for Foreign Affairs and defence experts of the different Specialist Commissions are all represented here.

The Inspection Commission is responsible for ensuring that the *JNICT* abides by the legislation in force (in fiscal matters as well as that related to science and technology) when carrying out its actions.

The following three institutions below, were set up under the auspices of the Ministry for Science and Technology. Given that they are of recent creation, their performances to date are limited.

Science and Technology Foundation

The Science and Technology Foundation (*Fundação para a Ciência e a Tecnologia- FCT*) channels the funding of programmes and institutions related to scientific research. It is responsible for the strengthening of infrastructure and the actions of training of research personnel through study agencies, both within the country and abroad. The *FCT* establishes protocols and agreements with other institutions dedicated to scientific research and technological development. Special emphasis is given to the dissemination of scientific results obtained, with the objective of increasing interest in these activities.

Science and Technology Observatory

By order of the Ministry for Science and Technology, the mission of the Science and Technology Observatory (*Observatório das Ciências e das Tecnologias- OCT*) is to ensure that the socio-economic agents have access to the results of scientific research. It also maintains an up to date inventory of national research resources, both physical and human. This new organisation has been granted certain privileges, to enable it to respond to the functions assigned. Examples of such functions are: support of the preparation of the annual R&D budget, collaboration in the elaboration of national plans and carrying out studies and analysis for the better planning of actions.

Institute of International Scientific and Technological Cooperation

One of the tasks of the Ministry for Science and Technology is the coordination of scientific and technological cooperation within the area of bilateral and multilateral cooperation agreements, especially those developed in the framework of the European Union, and assuring the collaboration and support of the competent services from the Ministry for Foreign Affairs. The Institute of International Scientific and Technological Cooperation (*Instituto de Cooperação Científica e*

Tecnológica Internacional- ICCTI) was created in order to carry out this mandate of the Organic Law of the Ministry for Science and Technology.

> *The Institute of International Scientific and Technological Cooperation is the institution in charge of directing, guiding and coordinating cooperation agreements in the area of science and technology, mobilising national and EU funds for such reasons, and without prejudice to the tasks assigned to the Science and Technology Foundation. (Article 9 - Law 144/96)*

This Law grants the Institute support functions to the Ministry for Science and Technology in the following activities:

- Activities inherent to Portuguese participation in EU programmes.
- Portuguese representation in international organisations promoting multilateral cooperation in science and technology.
- Coordination of bilateral and multilateral relations with other countries, in collaboration with the Ministry of Foreign Affairs and with other departments of public administration competent in European topics and foreign affairs.
- The Institute has complete administrative and financial independence.

Coordinating Cabinet for Scientific and Technological Policies

The Coordinating Cabinet for Scientific and Technological Policies (*Gabinete Coordinador da Política Científica e Tecnológica*) advises the Minister for Science and Technology on the coordination of activities within the framework of tasks of the Ministry for Science and Technology. This Cabinet, presided over by the Minister, consists of those responsible from public entities in charge of the development of scientific research; Science and Technology Observatory, Science and Technology Foundation, Institute for International Scientific and Technological Cooperation. It also supervises the work carried out in other organisations, such as universities, state laboratories, associated laboratories, the innovation agency and other public and private S&T entities.

Specialist Associations

The recently created Specialist Associations (26 August 1996) (*Colégios da Especialidade*) promote the participation of the scientific community in defining the policies for technological development to be followed in Portugal. The associations serve as an advisory organ to the Ministry of Science and Technology by bringing together teachers and researchers working in a specific area of science. The reason for the creation of these organisations is to provide for the future reform of the actual Higher Council for Science and Technology, granting the scientific community a higher degree of participation.

PUBLIC RESEARCH CENTRES
The Role of the Public Sector

Throughout the 70's, the more industrialised countries of the world in general confronted the successive reform of their national S&T systems. The reason for this is to be found in the increasingly important role of scientific research in the entire socio-economic context.

Although public administration is the only entity capable of mobilising a sufficient quantity of resources toward high-priority areas of greater interest, excessive dependence on government investment, (the case of Portugal) implies scarce autonomy and flexibility when designing and implementing the S&T system. Certain fundamental incompatibility exists between public financing and its manner of operating, and the inherent nature of scientific research.

In the case of Portugal, no solution to this problem has been found. The Portuguese government in its desire to make research promotion and development more flexible, created non-profit organisations to serve as driving-forces and to channel the S&T system. Initially, this idea is not entirely incorrect, but has shown itself to be insufficient in achieving its objectives. In addition, as the country is presently immersed in public spending cutbacks (to bring about convergence with the other EU Member States) the freezing of public investment has increased the inflexibility of public action in R&D.

The role of the State, through its public research centres, does not only consist of developing a series of R&D related programmes. Other activities are developed in public research centres which are related to technique rather than to the development of science and technology, such as: renovation of infrastructure; information on the national S&T system; development of technical standards; realisation of homologation, calibrations, and inspection activities; and even transformation of some scientific achievements in industrial applications. All this explains why the overall budget of a public research centre is usually higher than the quantity dedicated to "pure" scientific research; in some cases, it can even be up to six times greater.

The high-priority objectives of public research centres can be summarised in a series of fundamental points:

- Analysis of the physical characteristics and natural resources of the country (geology, agriculture, oceanographic characteristics ...).
- Analysis of human impact on nature and the consequences for the economy and society in general.
- Development of R&D activities in support of sectorial needs.
- Development of R&D activities aimed at overall national modernisation.
- International cooperation in scientific research, especially with Portuguese speaking countries.

In Portugal, there are 20 public research centres developing activities covering most of the range of scientific expertise. An exhaustive analysis of each is not possible, given that it would lead away from our primary objective, although reference will be made to the activities developed in the most important. Table 2-1 gives a list of the most important Portuguese public research centres.

Given that in Portugal as in most European countries, an important percentage of scientific research is carried out in universities, a brief analysis of this sector in Portugal will be given, as well as that of the public research centres, and the role played by the *JNICT* in their operation.

Table 2-1: Main Public Research Centres (Budget 1990-92) Millions of escudos *

Public Research Centres	1990	1991	1992
INETI	4,227	4,286.0	3,842.0
INIA	4,721	4,824.0	5,013
LNEC	1,870	2,984.0	3,350.0
INIC	2,840	2,471.5	2,012.9
CTQB	233	1,922.9	1,374.3
INIP	1,125	1,156.7	1,174.5
IICT	972	1,202.1	1,023.9
MDN	254	561.4	731.7
DGGM	777	547.0	686.0
CNCDP	—	977.3	684.0
INSA	322	579.5	598.9
IH	737	486.3	518.3
INMG	315	144.0	315.4
LNIV	232	203.8	276.5
CNIG	103	135.5	165.8
ENSP	—	63.7	106.4
OAL	—	21.9	95.3
DGQA	96	83.8	84.2
Total	18,975	22,651.9	22,052.4

Source: JNICT, 1992

Universities

Most of human potential in research is concentrated within the university environment in Portugal. According to OECD data, more than 50% of Portuguese scientists develop their activity inside the framework of these entities. Apart from their educational tasks, university centres develop a very important percentage of national scientific projects. Support to the different socio-ecnomic sectors through research in these centres has undergone substantial growth in recent years. Portugal has 33 public universities, including vocational technical schools.

OECD statistics further show that R&D spending in 1992 in universities was over 19 billion escudos, a figure representing more than 28% of total national

* Escudo. Average exchange rate in 1995: 110 Escudo = 1 $. 104 Escudo = 1 Deutschmark

R&D investment. In recent years, this level has increased to values of around 30%. The budgetary outlay of activities developed can be observed in Figure 2-2.

Figure 2-2: Distribution of R&D Spending in Higher Education

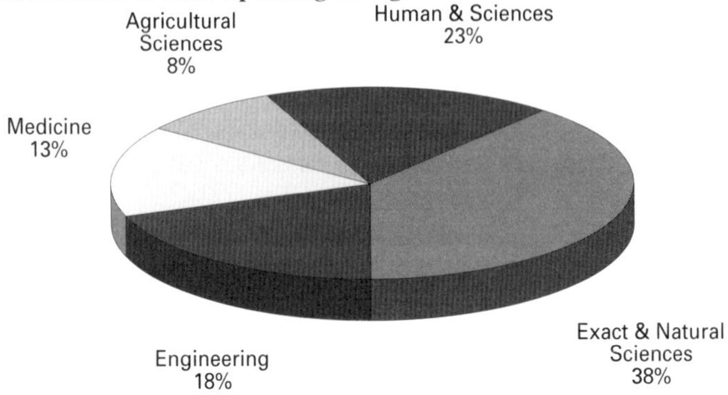

- Agricultural Sciences 8%
- Human & Sciences 23%
- Medicine 13%
- Engineering 18%
- Exact & Natural Sciences 38%

Source: OECD, 1992

Research sectors related to social sciences, and exact and natural sciences monopolise more than 60% of the budget, while the remaining sectors, perhaps of greater practical application due to their economic interest, spend only 40%. It can be deduced from this data, that university research may not be appropriately oriented. Taking into account that Portugal is in a process of industrial modernisation, it is certainly contradictory that only 18% of the total budget is dedicated to industrial and technological affairs. Evolution of this data has been positive in recent years, given that in 1980, this last percentage was only 13%.

Figure 2-3: Distribution of University R&D Spending by Socio-Economic Objective

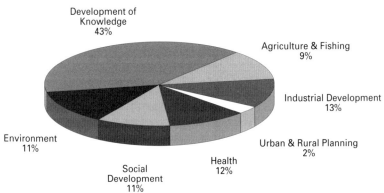

- Development of Knowledge 43%
- Agriculture & Fishing 9%
- Industrial Development 13%
- Urban & Rural Planning 2%
- Health 12%
- Social Development 11%
- Environment 11%

Source: OECD, 1992

With regard to the socio-economic objectives of research projects, the tendency corresponds to that of funds distributed according to activity.

Description of Main Public Research Centres

Ministry of Agriculture

The Ministry of Agriculture manages two institutes: The National Agricultural Research Institute (*Instituto Nacional de Investigação Agrária - INIA*) is responsible for research in topics related to agriculture, while the National Veterinary Research Laboratory (*Laboratório Nacional de Investigação Veterinária - LNIV*) is concerned with veterinary research.

National Agricultural Research Institute

The National Agricultural Research Institute (*Instituto Nacional de Investigação Agrária - INIA*) is one of the most important public research centres. Created in 1974, it has undergone important restructuring since then. The most important was in 1988, where it was granted a high degree of financial and administrative autonomy. The basic objectives of this institute are:

- To promote, coordinate and supervise all scientific activities related to agriculture.
- To elaborate proposals to the Ministry of Agriculture with the objective of promoting and developing R&D activities.
- To elaborate specific training plans for researchers in coordination with other public research centres.
- To disseminate and put in practice the results obtained through projects.

In 1993, personnel in this centre were over 2,600, of which approximately 250 were researchers, 90 assistants and more than 100 researchers in work placements.

This institute has a series of support departments to enable it to carry out its activities: Planning and Administration, S&T Information, Division of Statistical Analysis and Training Centre. Another series of departments also exist to coordinate the activities of the different institutes or laboratories associated to the *INIA*. These institutes are 8 in total and specialise in the following: agriculture, zoology, forestry, improvements in plantations, viticulture, fruit trees, production technology and analysis of land. Three more departments exist which are specialised in such concrete areas as: hydric resources; horticulture; social and rural economic studies.

The *INIA* monopolises almost all scientific research related to agriculture and livestock. Although the activities of this institute are wide, much still remains to be done. It should not be forgotten that the infrastructure and resources dedicated to agriculture in Portugal are scarce. These deficiencies can be summarised in three points:

- The level of cattle production in Portugal is clearly inferior to that of the other EU countries. (Half of Spanish or Greek production, for example)
- Investment is relatively low. In 1992, Spain or Greece (countries with long traditions in agriculture and livestock) dedicated nearly 15% of their GDP to these activities, while Portugal dedicated only 5%.
- The educational level of workers in this sector is not very high, and is below the European average.
- Both agricultural and livestock holdings are small, impeding concentration of effort.

Initially, the role of the *INIA* should be that of a driving-force and catalyst of resources used in the development of agricultural activities, but at present the Portuguese economy in general is undergoing rapid modernisation, in particular industry and services, leaving aside this important facet of the primary sector. Such a statement can be corroborated by the fact that the *INIA* budgets have been gradually reduced in recent years (in constant prices), while the incorporation of new human research capital has been slow, thus raising the average age of researchers above that of other Portuguese scientific institutes.

The *INIA* budgets can be seen in Table 2-2:

Table 2-2: *INIA* Budget (1989-1992). Current Prices. Millions of Escudos

	INIA Budget Spending on Personnel	Others	Self-financing	Public Funds	Total
1989	3,228.3	337.9	817.7	453.4	4,837.2
1990	3,371.5	180.7	788.3	381.1	4,721.6
1991	3,691.4	61.8	677.2	393.9	4,824.4
1992	4,010.3	50.2	560.4	392.1	5,013.0

Source: INIA

National Veterinary Research Laboratory

The National Veterinary Research Laboratory *(Laboratório Nacional de Investigação Veterinária - LNIV)* develops R&D projects related to the health of livestock, improvement of breeds, conservation of food, food production, etc. The legislative framework grants it a series of competencies;

- Establishment of standards, quality control and regulations.
- Development of R&D activities characteristic of these sectors.
- Quality control, analysis, etc.

The laboratory consists of six departments (pathology; bacteriology and virology; parasitology; food sciences; biochemistry; biological and therapeutic products). Activities are developed in six centres spread throughout the country (Mirandela, Oporto, Viseu, Castelo Branco, Evora and Faro).

Total staff is 400 and the budget in 1992 was more than 270 million escudos.

Ministry of Industry and Energy

National Engineering and Industrial Technology Institute.

This institution was created under the name of National Engineering and Industrial Technology Laboratory (*Laboratório Nacional de Engenharia e Tecnologia Industrial - LNETI*) in 1977 (Decree-Law 548). Since its foundation, this entity has undergone diverse restructuring, the last of which took place in 1992, resulting in its current name of *INETI* (*Instituto Nacional de Engenharia e Tecnologia Industrial*). Its constitution was the result of the fusion of two of Portugal's most renowned research centres: The National Industrial Research Institute and the Nuclear Energy Agency.

In the 80's, the Council of Ministers introduced the ten year long Plan of Technological and Industrial Development of the Production Sector. It is in this context that *INETI* has developed most of its activities.

In accordance with its own statutes and the legal framework regulating its performances, the objectives of *INETI* are the following:

- To promote and implant R&D activities in industry and energy sectors.
- To collaborate in the application of industrial and energy policy.
- To contribute to national technological and industrial development by improving the competitiveness of Portuguese enterprise.
- To provide assistance for the development of new industries and products.
- To promote the diffusion of scientific research results.
- To coordinate training projects in areas related to scientific research.

Projects presently being developed in this Institute are related to diverse scientific areas, the most noteworthy of which are: information technology, production technology, energy technology, environmental technology, biotechnology, fine chemistry, food, nuclear science and engineering.

Of the 824 staff, 275 are research personnel.

Institute for Geology and Mining

The Institute for Geology and Mining (IGM) depends on the Ministry for Industry and Energy and was created in 1993 as a result of the restructuring of the now extinct Directorate General for Geology and Mining. The Institute is responsible for geologic research within the national boundaries and the continental platform. Studies are also carried out in the use of national geological resources, advising mining industries established in Portugal.

At present, the Institute employs 19 researchers, of a total staff of 78.

Ministry of Health

National Health Institute "Dr. Ricardo Jorge"

Both the Ministry of Health and regional health services can avail of this institute as scientific-technical back up. Activities carried out here are related to health, clinical, biomedical, and pharmaceutical-biological R&D as well as food and environmental safety. The Institute: coordinates the activities of various national research centres; promotes health R&D through research units in the national health system; is responsible for the training of personnel, especially in areas related to health administration and public health; and is responsible for epidemic surveillance in collaboration with the national epidemiology centres .

The percentage of research personnel over total staff is somewhat low at present; of a total of 339, research personnel are 27.

Ministry of Planning and Administration

Institute for Scientific Tropical Research

The Institute for Scientific Tropical Research *(Instituto de Investigação Científica Tropical - IICT)* comes under the aegis of the Ministry of Planning and Administration and its origins go back (although under a different name) to 1883. Its activities are deeply rooted in Portuguese research. The main objective is the promotion of scientific research and techniques in tropical regions, cooperating with other countries in these areas. It has 23 specialised research centres, which in turn, are integrated into six scientific departments covering the areas of research to be carried out: land sciences, geographical engineering and biological sciences, agricultural sciences, historical sciences, economic and social sciences, ethnological and ethno-museological sciences. It also has a centre for scientific-technical documentation and information on tropical regions. The reserves of the centre are some of the most important world wide in this field.

The fact that this Institute employs a total of 500 people, 200 of whom are researchers and university teachers with considerable experience in tropical research, gives some idea of the size and scope of the research carried out here.

Ministry of Environment and Natural Resources

National Meteorology Institute

The National Meteorology Institute *(Instituto Nacional de Meteorologia - INM)* depends on the Ministry of Environment and Natural Resources, although its functions are developed in close collaboration with the National Meteorological Service, created in 1946.

Within its present areas of responsibility are the study of meteorology, air quality and seismology. Areas of research being developed are fundamentally concerned with the composition of the atmosphere, the greenhouse effect, renewable energy, prevention and prediction of natural catastrophes and the impact of the environment on human health and diverse socio-economic activities.

Of the 50 staff, 20 are research personnel.

PLANNING OF SCIENTIFIC RESEARCH: MAIN PUBLIC PROGRAMMES

Portugal has experienced profound changes since the end of the 80's in the planning of scientific research. Previous to this date, concrete plans of action which oriented scientific research toward the most important socio-economic areas in Portugal were practically non-existent. Public administration limited its activities to subsidising public research centres so that each one acted in line with its own criteria. This situation resulted in a clear disorientation in the objectives to be developed, duplicity of effort and administrative malfunctions. In general, projects carried out did not fulfil their function of dynamism of Portuguese society, and in particular, of production. It should not be forgotten that one of the fundamental objectives of all S&T systems is the development of new processes or products that can be transferred immediately to the industrial sector of the country, with the objective of endowing it with competitive advantages as regards third parties. Without adequate development, planning and coordination, such focused concentration of the main part of scientific production is not possible.

It was at the end of the 80's, when a series of programmes dedicated to making the S&T system cohesive was introduced in Portugal for the first time. Much has been achieved since then, due to the impulse of the European Union, and the more or less definitive restructuring of public organisations in charge of research management. Scientific policy in Portugal is appropriately orientated at present, due to the development of specific R&D plans, the most important of which are described below.

Programme for Educational Development in Portugal

The Programme for Educational Development in Portugal (*Programa de Desenvolvimento Educativo para Portugal - PRODEP*) was initiated in 1990. Its fundamental objective is the improvement of Portuguese education in its scientific and technological aspects. The Cabinet of Studies and Planning was responsible for development at the theoretical level. Most noteworthy of its initial intentions was: "the generalisation of the Portuguese educational system in order to situate it at a level similar level to that of its European partners, in such a way that: regional imbalances are eliminated and asymmetries between academic and vocational education corrected; the educational infrastructure is modernised in order to train new researchers capable of satisfying the technological and scientific needs of the nation; educational quality is improved by promoting academic success."

The objectives of *PRODEP* are not limited to actions at higher level education. The field of action also covers the pre-university level, impelling students to orient their career towards science and technology.

In the area of strictly third-level education, main lines of action consist of: improvement of university infrastructures, especially in polytechnics; improvement in teacher training; and the promotion of universities with lower levels of development, mainly those to be found in the north of the country.

PRODEP contemplates the following 5 sub-programmes (Table 2-3):

Table 2-3: *PRODEP*

Sub-Programme	Budget (Millions of ECUs)
Equipment	362
Vocational education	194
Post university Education	71
University education	319
Implementation and Evaluation of the Programme	4

Source : Portuguese Ministry of Science & Technology.

The development of university education was established as a high-priority objective, in an attempt to increase human research capital in the long term with which to supply the Portuguese S&T system. Concrete actions are carried out with the objective on increasing the number of students entering into technological careers. Specific plans are also created to improve teacher training.

Both plans and budgets are reflected in Figure 2-4:

Figure 2-4: Plans and Budgets. University sector

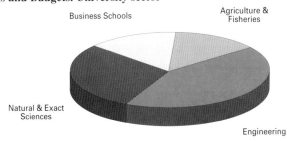

TOTAL BUDGET: 319 Million ECUs
Source: Portuguese Ministry of Science &Technology

Total Budget: 319 Million ECUs

Source: Portuguese Ministry of Science & Technology

Specific Programme for the Development of Portuguese Agriculture

The high-priority objective of *PEDAP* (*Programa Específico de Desenvolvimento da Agricultura Portuguesa* - Specific Programme for the Development of Portuguese Agriculture) is the development of Portuguese agriculture and livestock. It was approved in 1985 and its term of execution was up until 1995. Even from the outset, the programme not only concentrated on the agrarian R&D, but also contributed to the development of this entire economic sector. Given the date of initiation, *PEDAP* was one of the first programmes with heavy EU financing. Almost 78% of its total budget was provided by the European Commission, while the remaining 22% was contributed by the Portuguese Government.

The programme is made up of 18 national and regional programmes, 3 of which deal exclusively with agricultural and livestock research. Projects are shown in Table 2-4.

Table 2-4: *PEDAP* Projects

Construction, equipping and computerisation of vocational training centres for farmers and livestock breeders.
Creation of training centres for specialists and advisors in agricultural matters. Construction and equipping of the Centre for Chemical and Biological Technology.
Realisation of cartographic studies for the definition of the most effective structural alternatives for the development of national agriculture.

Source: Portuguese Ministry for Science & Technology

Specific Programme for the Development of Portuguese Industry

PEDIP (Specific Programme for the Development of Portuguese Industry - *Programa Específico de Desenvolvimento da Indústria Portuguesa*) is perhaps the most ambitious scientific programme carried out under the auspices of the European Commission since Portugal's incorporation in the EU. It was initially approved for a four year period in 1988, at the end of which, the second phase of the programme (*PEDIP* II) began and is presently in force. The fundamental objectives have been the following since the initiation of the Programme:

- To revitalise the Portuguese industrial base.
- To promote the creation of new industries with greater technological potential.
- To eliminate the structural disadvantages of Portuguese industry in relation to competitors, bringing about convergence with the EU industry average.

The areas of research within this Programme are as follows:

Programme 1: Infrastructure.

- Basic infrastructures, energy, communication infrastructure.
- Technological infrastructure, metrology laboratories, sectorial technological centres, institutes for the dissemination of new technologies, technology transfer centres.

Programme 2: Training

- Vocational training, training of trainers, specialists and technicians, promotion and development of pedagogic methods, courses in administration, reinforcement of relations between organisations responsible for continuous training and enterprise.

Programme 3: Initiatives for production and investment

- System of initiatives *PEDIP* (*SINPEDIP*); investment in R&D; innovation and modernisation, production quality, industrial design and environmental protection.
- Measures for the rationalisation of energy consumption.
- Modernisation of industrial sectors.

Programme 4: Financial aid

- Financial engineering; financial aid for investment in research projects, creation of venture capital societies, capitalisation of industries in decline.

Programme 5: Production initiatives

- Initiatives dedicated to the improvement of production by enterprise, support to sectorial collaborations, protection of research results.

Programme 6: Industrial quality

- Quality and industrial design, development and improvement of standards, certification of processes, initiatives for improvement and quality.

Programme 7: Measures for the dissemination of research projects

Programme 8: Strategic integrated projects. Sectorial programmes.

- Programmes for the Integration of Technology and Electronics. (*PITIE*)
- Programme for the development of equipment. (*PROBIDE*)

Such an extremely ambitious project contemplates all types of action, both sectorial and intersectorial, and therefore called for the creation of a complex lattice of horizontal and vertical programmes and sub-programmes, capable of responding to such an important challenge.

The main part of the programme is developed under the direction of the Ministry of Industry and Energy through the *PEDIP* Administrative Cabinet. Projects are carried out in the public research centres (*INETI, IPQ, IAPMAI*, etc.) in collaboration with the different business sectors.

The *CIENCIA* Programme

The programme for the Development of National Infrastructure for Science, Research and Development - *Criação de Infraestruturas Nacionais de Ciência, Investigação e Desenvolvimento - CIENCIA*) began in 1990 and although it ended in 1994, it is still of importance, given that it laid down the basis for the future development of the Portuguese S&T system. Contrary to the previous programmes, this one is exclusively scientific and technological.

The fundamental objectives can be summarised in the following three :

- To strengthen the scientific and technological potential of the country.
- To improve scientific infrastructure.
- To reduce regional imbalances in the Portuguese socio-economic environment.

CIENCIA was mainly financed by the European Union and its budget was 253 million ECUs in 1994. Participation by the Portuguese government was reduced in the last years of the Programme, given their desire to reduce the high level of public deficit, which in 1994 was limited to 36%.

The first of the high-priority objectives was concerned with the following: the development of scientific disciplines in areas of scientific knowledge which require a high level of multi-sectorial coordination; (information technologies, production and energy technologies, biotechnology, etc.); those linked to the exploitation of natural resources in Portugal (agriculture, fisheries, mining); or those which require coordination of a large number of human research capital (health). The second objective refers to the improvement of infrastructure in research centres, public and private (public laboratories, equipment, communication networks, technology parks, etc.). The third objective, and perhaps one of the most important, refers to the cohesion that must be achieved among Portuguese regions regarding scientific research. As in the majority of countries of this context, scientific potential is almost exclusively limited to areas of greater development, and in particular, to the environment of large urban nuclei. In the case of Portugal, the region of Lisbon is where there is greater intensity of concentration. In *CIENCIA*, contributions to this region were limited to only 50% of the total, while the regions of Madeira and Alentejo were specially favoured in the assignment of financial resources.

Within these three basic lines of action, special emphasis was given to the need to endow Portugal with an appropriate number of researchers, a fundamental element for the convergence of the Portuguese S&T system with that of the European Union. Specific training initiatives were developed with the initial objective of training 2,600 new researchers. This series of actions may have been the most successful, given that initial financing received was not reduced in the final years, but gradually increased.

At the express desire of the European Commission, *CIENCIA* also included a strong transnational component, to enable synergy to be established between Portuguese and European research centres, and the participation of national companies in concrete collaboration agreements with the main production sectors in European common space.

Table 2-5: *CIENCIA*

PROGRAMMES
Development of infrastructure in high-priority areas
Information and Telecommunication Technologies
Production and Energy Technologies
New Materials
Medical Sciences and Technologies
Agricultural Sciences
Biotechnology and Pure Chemistry
Marine Science and Technologies
Strengthening of Infrastructure in High-Priority Areas
Advanced training and innovation in high-priority areas
Training in High-Priority Areas
Technology Parks
Support to R&D infrastructure
Exact Sciences and Engineering
Environmental Sciences
Economy and Management
Infrastructures for Common Use
Infrastructure for the Dissemination of Scientific Results
Training of Scientific Personnel
Implementation programme
Technical Assistance
Support to Programme Management

Source: OECD, 1994

The *JNICT* has been responsible for the management of *CIENCIA* since its initiation. In order to carry out its tasks, specific advisory committees were set up, of which the following are the most outstanding: National Council for the *CIEN-*

CIA programme; the Support Committee; Regional Committees; and the National Committee of Experts.

STRIDE

STRIDE (Science and Technology for Regional Innovation and Development in Europe) is an EU programme which provides support to operative, national S&T programmes. The European Commission started this initiative in 1990, with the objective of achieving the convergence of the S&T systems of the less favoured EU countries; the entire national territories of Greece, Portugal and Ireland and certain regions of the other Member States. Financing is provided by the European Structural Funds and the individual governments involved in actions.

The first phase of STRIDE was three years (1990-1993) and financing was greater than 70 million ECU. The second phase began in 1993, for another period of three years. Basically, the programme is financed by the European Commission through the European Regional Development Funds (ERDF), the main objective of which is to promote EU inter-territorial cohesion. Remaining funding comes from the Portuguese State.

Again, the *JNICT* was responsible for the development of this programme, and set up various advisory committees and agents. The most important areas of research are reflected in Table 2-6.

Table 2-6: STRIDE Programme

INTERNATIONALISATION OF THE S&T SYSTEM
Participation in international programmes
Internationalisation for balanced development
STRENGTHENING OF NATIONAL, TECHNOLOGICAL CAPACITIES
Innovation Agency
Consortium for the development of new technologies
DIVERSIFICATION OF THE PRODUCTION SYSTEM
TECHNICAL ASSISTANCE, ADMINISTRATION AND EVALUATION

Source: JNICT

Through such initiatives, the European Commission aimed to:

- Reinforce the capacity of the Portuguese S&T system in order to establish appropriate national participation in organisations related to EU research.
- Vitalise Portuguese participation in the R&D Framework Programmes of the European Union.
- Stimulate R&D in lesser developed regions of Portugal.
- Promote research in Portuguese enterprise through specific contracts and technology transfer.
- Promote cooperation among the agents involved in scientific research, especially between public research institutes and enterprise.

PRAXIS XXI

One of the most recent R&D programmes in Portugal is that of *Praxis XXI*. Its main objective is the development of national scientific research, mainly through agreements between public research centres and enterprise.

The programme is financed by the Community Support Framework of the European Commission and the Portuguese State, as well as enterprise involved in the projects. Management and administration of the programme falls on the Ministry of Science and Technology.

R&D activities included are those which demand cooperation between enterprise and public research centres, associated by means of a consortium contract. Among the concrete objectives to be fulfilled are :

- To support to generic or horizontal technologies, through sectorial actions, permitting the creation of new products, services and processes.
- To support the participation of national consortia in concerted actions of research and technological development in European or transnational programmes.
- To integrate training activities associated to research and technological development into consultancy actions in the projects.

Programme actions aimed at increasing the degree of competitiveness of companies or groups of companies usually include :

- Actions of pre-competitive research for the development of new technologies and "know how."
- Training actions usually through a series of job data bases framed within the initiative "Mobility between the S&T System and Enterprise."
- Consultancy actions for the development of the technological context of projects, and the dissemination and appraisal of the obtained results.
- Actions for the promotion of economic results of scientific research.
- Prototypes and tests dedicated to validating products or processes developed at the laboratory level in enterprise.
- Technological transfer of results, facilitating their adoption and adaptation in public research centres and the production sector as a whole, with the objective of improving competitiveness.

As is logical, one of the aspects which limits scientific research in the business production sector, is the investment risk that must be taken by the company. This is one of the reasons why *Praxis XXI* concentrates an important part of its financial activities on high risk investment; those areas of science related to basic or pre-competitive research. In these cases, funds provided by the Programme may cover almost the entire project.

Given that from the very beginning of the programme, some type of organisation was necessary to carry out administrative tasks in collaboration with the

Ministry of Science and Technology, the Administration Cabinet of the *Praxis XXI* Programme was created. In addition to this role, the Cabinet also supervises and advises on the programmes to be carried out. Indeed, it was set up as an interface between the production sector and the funds assigned to this initiative.

More significant projects to be developed within the framework of *Praxis XXI* are highlighted in Table 2-7.

Table 2-7: The *PRAXIS XXI* Programme

Promotion of the Base of the S&T System
Technological Development Projects of Small and Medium Dimension in Social and Human Sciences
Applied Biology
Biotechnology
Health Sciences
Consortium Projects
Exact Sciences : Areas of Mathematics, Physics and Chemistry
Natural Sciences.

Source: Portuguese Ministry of Science & Technology

Training actions related to the realisation of R&D projects are financed by the European Commission and are parallel to the above. Such actions last a maximum of three years.

Table 2-8 shows the number of projects and their financing levels for the period 1993-1996. The growth in the number of projects has been substantial. The reason for this is due to the fact that public administration gave new drive to this action, paralysed at the outset of the programme regarding the granting of aid. New legislation and procedures for the presentation of candidates resolved the blockage that froze this initiative in the first years of its existence.

Table 2-8: Projects in Progress and Financed by the Ministry for Science & Technology *PRAXIS XXI*

Year	1993	1994	1995	1996	Total
Financing (Millions of escudos)	2,061	2,071	2,330	12,720	19,182
N° of Projects	94	16	220	857	1,187

Source: Portuguese Ministry of Science and Technology, 1997

A short evaluation of the state of the Portuguese S&T system shows that the main R&D priorities are covered by the initiatives of scientific and technological development plans at the national level. In spite of everything, and according to OECD published reports, these do not show the sufficient degree of detail and precision required; they should be more specific when determining the strengths and weaknesses of the Portuguese technological environment, defining in particular, high-priority areas of research, vital to the national socio-economic area.

It is still too early to evaluate the new restructuring of public organisations at ministerial level responsible for the development of scientific research in Portugal, although it can be confirmed that previous entities did not formulate a specific, national plan for R&D which brought together national institutions (educational system, public research centres, enterprise, etc.). This plan should be based on a precise strategy, capable of responding to the entire set of national, technological needs, long and short term, assuring Portuguese participation in the common path being followed by the European countries.

To date, *CIENCIA* and *PEDIP* have been the programmes which have come closest to fulfilling their established objectives. *CIENCIA*, administrated by the JNICT, was designed to stimulate pre-competitive research and national technological development, through the improvement of infrastructure and the training of human research capital. *PEDIP* was established to improve the productive capacity of enterprise, searching for the appropriate level of competitiveness. The problem was that both programmes were based to a great extent on outside financial support, the European Union in particular. This can in no way substitute adequate investment by public administration.

As for the more recent *Praxis XXI*, the need to clarify strategies, make the programme more transparent, and cohesive in its socio-economic objectives, are all issues raised by the scientific community.

It should again be stressed that any S&T system, in order to be effective, need not be centralised, nor should it be excessively rigid. The existence of common objectives and strategies for the country as a whole, as well as ease of adaptation to the concrete requirements of each region and sector is necessary.

The recent creation of a ministry in charge of the coordination of public R&D activities will probably be one of the solutions to the problem outlined, should it be able to design and implement national R&D plans (assuring personnel mobility, eliminating bureaucratic obstacles and encouraging a progressive decentralisation of scientific research). Likewise, it will be capable of managing the budget dedicated to such activities, promoting the convergence of the Portuguese S&T system with that of its European partners.

TECHNOLOGICAL DEVELOPMENT AND INNOVATION IN THE PORTUGUESE BUSINESS SECTOR

The evolution of the Portuguese economy has evolved in a similar manner to that of its EU neighbours in the last 25 years, becoming more and more service based in detriment of agriculture, livestock and fisheries. Most noteworthy in the recent evolution of the Portuguese productive sector has been its capacity to maintain a considerable level of exports. Since Portugal's entry into the European Union in 1986, the presence of both the industrial and services sector in the Single Market has been growing, especially through the financial and textile sub-sectors. On the other hand, the inability of the agricultural and fishing sectors to adopt and adapt to new technological and innovative changes has forced them to almost abandon the market, in benefit of other countries such as Spain or France.

Traditionally within the productive sector, it was the textile sector which showed more important rates of activity. In the last few years, competition with the emerging economies of eastern Asian have made it necessary to implement technological changes in order to maintain their competitiveness. Portuguese companies are also strong in those sectors of production where they have been able to establish themselves: paper, plastic, chemical, shipbuilding etc. Proven political stability, accompanied by low labour costs compared to those of their EU partners, have led to the establishment of companies with high technological content in the country, especially in the motor sector.

Portugal is one of the poorest countries in the European Union. GDP is about 40% of the EU average, although the standard of living in the main urban nuclei approaches that of the other EU countries. The services and industrial sectors have been established around such zones, especially in the metropolitan area of Lisbon, in the Tajo Valley and in the area surrounding Oporto. But the agricultural area of Alentejo, in the south of the country, lacks virtually any industry and GDP per capita is only 20% of the EU average.

In view of all this, perhaps the most outstanding single fact is that an important productive sector does indeed exist, but lacks a sufficient technology level.

Recently, Portuguese industry has been highly intensive in financial capital regarding business (very high interest rates), and in human resources regarding production. This suggests that investment carried out in the development of new technological processes is scarce, reasonable perhaps in the context of public administration given that Portugal does not avail of the means to realise such investments, although its occurrence is less logical in the enterprise sector as it is not so weak. Another aspect to be pointed out, in contra the Portuguese S&T system, is its excessive dependence of the acquisition of foreign technology due to scarce domestic research.

Before going into the basic parameters marking the technological development of Portuguese industry, the fact should be mentioned that a high number of companies are very small (more than 90% have less than 4 workers), and almost all are SMEs (more than 99% have less than 50 employees). Taking into account the type of company developing its activities in Portugal, it is easy to understand why they depend to such an extent on the acquisition of foreign technology.

The main divisions of spending carried out in enterprise regarding R&D can be observed in Table 2-9, taking into account the type of research, activity and source of financing. Unfortunately more recent public authority statistics on this aspect are not available, which is understandable given the diversity and dispersion of expenditure carried out in the industrial sector (it is not centralised, which would be the case in public administration).

Table 2-9: R&D Spending in Industry
            Millions of Escudos (1990)

	Spending	%
Research		
Basic Research	210.6	1.6
Applied Research	3,113.5	22.9
Experimental Development	10,261.5	75.5
Type of Economic Activity		
Agriculture and Fisheries	1.3	—
Mining	109.0	0.8
Manufacturing	9,613.9	70.8
Services	3,861.4	28.4
Source of Financing		
Public Administration	882.1	6.5
Own Funds	12,064.8	88.8
Private Non-Profit Institutions	50.8	0.4
Universities	6.0	—
Foreign	581.9	4.3
Type of Spending		
Current	9,906.0	72.9
Personal	6,351.9	46.7
Other Costs	3,554.1	26.2
Spending by Activity	3,679.6	27.1
Land and Buildings	766.3	5.6
Instrumental and Equipment	2,913.3	21.5
Total	13,585.6	100.0

Source : OECD, 1995.

As can be observed, industrial and services sectors are those which absorb greater quantities of resources, clearly demonstrating the supremacy of both sectors in the Portuguese economy, while others, such as mining, agriculture or fishing, make up less than 1% of total investment. Although such displacement of investment is occurring throughout all Europe, it is more prevalent in Portugal. In countries such as Spain or France, the competitiveness of the primary sector is being revitalised to make it more productive.

Another noteworthy aspect is the low level of public funding received by enterprise for the development of R&D activities. Of total investment, only 6.5% comes from public administration, while in other countries (Spain, France, Italy, etc.) it is about 20%.

An important factor which is not reflected in the table, is the type of company carrying out R&D activities. According to OECD reports published, little more than 15% of Portuguese companies admitted carrying out concrete tasks in this field, and only a small percentage have budgets assigned to this area. Such data is the result of a series of surveys carried out on a sample of 1,500 companies (Table 2-10).

Table 2-10: R&D Expenditure According to Company Size in 1990

	Companies	R&D Expenditure Millions escudos	%	Company Expenditure %
Production				
Less than 5,000 million escudos	81	1,774.1	13.0	21.9
5,000-10,000	30	2,798.2	20.6	93.3
10,000-25,000	35	2,075.1	15.3	59.3
25,000-50,000	19	1,510.7	11.1	79.5
50,000-100,000	15	2,661.0	19.6	177.4
100,000 or greater	14	2,766.5	20.4	197.6
Total		13,585.6	100.0	70.0
Number of Employees				
less than 100	57	1,097.0	8.1	19.2
100-250	36	2,518.6	18.5	70.0
250-500	34	1,470.2	10.8	43.2
500-1000	35	2,961.9	21.8	84.6
1,000-5,000	27	4,237.6	31.2	156.9
more than 5,000	5	1,300.3	9.6	260.1
Total	**194**	**13,585.6**	**100.0**	**70.0**

Source: OECD, 1994.

Table 2-11 shows R&D spending in Portuguese companies according to the type of activity developed. As can be observed, the services sector is acquiring greater importance and only the industrial sector as a whole is larger, which is logical given that it is also the biggest in size. Other such sectors as electricity, construction, etc. dedicate clearly inferior quantities to R&D.

Table 2-11: R&D Expenditure in Enterprise According to Activity (Millions of Escudos)

Sector	1988	1990	1992
AGRICULTURE	14.7	1.3	12.8
MINING	162.4	109.0	75.4
INDUSTRIAL	4,916.1	9,613.9	13,419.2
Food	326.0	472.8	696.3
Textile	192.0	368.6	400.8
Wood, paper	880.2	1,590.4	1,042.5
Petroleum, Chemical	1,312.7	2,006.3	2,379.8
Non-metal mineral products	29.4	164.0	220.8
Metals	41.4	162.6	495.6
Metal products	98.9	221.8	585.0
Machinery, equipment	2,035.5	4,627.4	7,568.2
Other manufacturing	—	—	30.2
Recycling	—	—	—
ELECTRICITY, GAS, HYDRO RESOURCES	574.2	148.3	221.3
CONSTRUCTION	6.5	12.2	15.6
SERVICES SECTOR	1,677.1	3,700.9	3,707.9
Sales, commerce	—	—	—
Hotels, Restaurants	—	—	—
Communications	683.0	1,488.4	2,113.2
Transport and storage	150.7	97.7	91.8
Financial intermediation	—	—	—
General financial activities	839.7	2,029.2	1,500.3
Communication, society	—	—	2.5
TOTAL	7,351.0	13,585.6	17,452.2

Source : OECD, 1995.

Figure 2-5: Growth in R&D Expenditure of the Main Sectors of Portuguese Enterprise (Millions of Escudos)

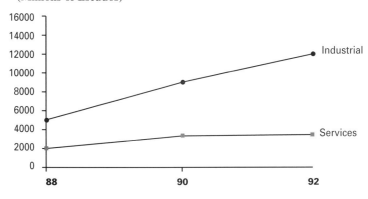

Source: OECD, 1995

Investment growth in the two main sectors of the Portuguese economy (services and industry) are illustrated in Figure 2-5.

The continuing lack of appropriate competitiveness in Portuguese industry in general forces the need for greater R&D investment. This also demonstrates the growing interest in the different areas of scientific research, in such a way that it can be stated that Portuguese industry has accepted the challenge of the competition established by the surrounding countries. This is not so in the case of the services sector. After rapid growth in R&D when the country joined the European Union, the sector as a whole was able to come up to the level required, and the investment level stabilised. This may seen optimistic, but experience shows that investment in technological and innovative development in general, being a dynamic element, must always be constant and growing, otherwise important market quotas will be lost.

As for the number of full time dedicated researchers in enterprise, the total in 1991 was not greater than 2,000. Initially, this figure cannot be considered high, especially if compared proportionally with that of other EU countries. A possible reason for this may be that most companies do not consider R&D to be even among the 12 main factors for innovation. This opinion was verified by the Cabinet of Studies and Planning of the Ministry of Industry (1992). Other factors were included within the survey: purchase of necessary equipment, improvement of production, adaptation of machinery, the pressure of competition, etc. R&D only reached first place in "less important" factors.

Table 2-12: R&D Personnel in Industry (In full time equivalent) 1990.

	Scientists	Technicians	Others	Total
Agriculture	—	0.2	—	0.2
Mining	3.2	11.7	15.2	30.1
Industry	210.6	938.8	319.3	1,468.7
Services	222.8	174.3	100.5	497.6
Total	436.6	1,125.0	435.0	1,996.6
Percentages	21.8	56.35	21.79	100.0

Source: OECD, 1995

According to the results of the same study, enterprise considered that R&D was important and that would somehow affect the future of the company; the problem was that this possible impact was only classified in ninth position within possible threats to innovation.

Data regarding total personnel involved in R&D activities in Portuguese enterprise and the capital dedicated to these activities can be seen in Table 2-13.

Table 2-13: R&D Personnel and Expenditure in the Production Sector Millions of Escudos

Sector	Personnel	Current Spending	Capital	Total
Agriculture	0.4	1.6	—	1.6
Mining	27.0	130.0	60.0	190.0
Food	264.0	731.6	77.1	808.7
Textile	525.2	540.9	23.5	564.4
Timber	27.6	48.2	0.5	48.7
Paper	184.9	529.4	51.1	580.5
Chemistry	594.1	1,467.9	129.0	1,596.9
Non metal mineral	86.1	115.4	36.1	151.5
Metal	192.8	1,028.0	24.6	1,052.6
Transport machinery	1,571.6	4,141.6	459.2	4,600.8
Electricity and gas	703.2	2,588.0	343.0	2,931.0
Construction	—	—	—	—
Transport & Communications	1,246.6	3,572.4	417.2	3,989.6
Services to Companies	183.2	1,360.7	35.7	1,396.4
Health	—	—	—	—
Research institutes	4.2	7.0	4.0	11.0
Total	5,610.9	16,262.7	1,661.0	17,923.7

Source : JNICT / SEFOR

Table 2-14: ERD - Acquisition of Technology & Value Added. (Millions of Escudos)

Sector	R&D Expenditure	Acquired Technology	Value Added	R&D as % of Value Added	Technology Acquired as % of Value Added
Food	823.1	473.0	55,390	0.85	1.49
Textile	14.0	368.4	32,372	1.14	0.04
Paper	549.3	1,492.6	52,100	2.87	1.05
Chemical	322.2	459.4	35,402	1.30	0.91
Other Chemical	977.7	1,423.6	28,868	4.93	3.39
Rubber products	88.3	25.1	1,014	2.48	8.70
Non electric machinery	160.0	659.7	14,682	4.49	1.09
Electric machinery	2,736.0	3,799.0	54,581	6.96	5.01
Electricity and Gas	191.7	148.2	297,354	0.05	0.06
Transport	2,358.0	97.7	52,011	0.19	4.50
Communications	—	1,488.4	197,016	0.76	—
Services	320.6	2,026.2	2.314	87.56	13.86
Other sectors	—	76.6	—	—	—
Total	7,400.0	9,400.0			

Source: JNICT

As previously pointed out, one of the problems of the production sector in Portugal is its dependence on technology acquired abroad in comparison with the scarce level of domestic research. (Table 2-14). In 1990 the total dedicated to R&D was 74 billion escudos compared to 94 billion dedicated to the acquisition of technology. The reason for this has already been indicated. Given that the majority of Portuguese companies are small, they fail to make concrete provisions dedicated to scientific research in their budgets, and therefore depend heavily on technology from abroad. Only some of the larger companies can afford this and are sensitive to inhouse elaboration of productive processes and innovative changes.

All indicators shown up to this point reflect a situation which could, in the long term, limit the competitiveness of certain Portuguese companies. In the last decade, the government designed and implemented support programmes for business innovation. The most important have already been outlined but reference will be made here again, especially to the *CIENCIA* programme which has promoted the introduction of integrated information systems in enterprise, and has improved energy technologies and production processes. For the first time not even the agricultural sector was left aside; specific R&D centres were established (*ICETA, ICAM, IISA*). It is also expected that the new restructuring in the administration of PRAXIS XXI, will bring about the fulfilment of its objective; to increase the degree of competitiveness of the Portuguese production sector.

According to information provided by the OECD, in order to encourage the productivity of the Portuguese socio-economic environment, the creation of a greater number of specific programmes or public research centres is not necessary, but rather, what is needed is the consolidation of their activities as well as their orientation toward high-priority, national interests.

PORTUGUESE PARTICIPATION IN INTERNATIONAL PROGRAMMES

Before 1986, the date of Portugal's incorporation to the European Union, the Portuguese scientific community was isolated from that of its neighbouring countries. Traditionally, research was carried out in universities or laboratories in foreign countries, and in general it can be stated that technology transfer was scarce. True internationalisation of Portuguese scientific research came about after the country's incorporation to various international organisations.

The NATO Science and Technology Programme

The first international programme of outstanding importance in which Portugal participated was the NATO Science and Technology Programme in 1970. Portugal was involved in the following projects:

Table 2-15: Portuguese Participation in EU Framework Programmes (Millions of Escudos)

Second Framework Programme	N° of Programmes with Portuguese Participation	EU Financing (Millions of escudos)
DOSES (statistics)	2	96
MONITOR/SAST	1	78
BRITE/EURAM	44	10,188
MAST	8	811
AIM	7	315
DRIVE	4	382
DELTA	9	495
ESPRIT II	74	13,484
RACE I	17	3,839
Radiological protection	3	146
STD II	15	1,211
Aeronautics	10	430
JOULE	22	1,741
Large Infrastructure	1	800
Recycling	25	1,115
BRIDGE	8	1,132
ECLAIR	7	1,182
FLAIR	4	540
STEP	19	1,934
EPOCH	9	491
VALUE	2	187
EUROTRA	2	752
BCR	16	311
Nuclear energy	2	3,065
Administration of Agricultural R&D	10	1,010
Total	321	45,735

Source: JNICT.

Table 2-16: Programmes of the III Framework Programme with Portuguese Participation

Information Technology
Communication Technology
Telematic Applications of Common Interest
Industrials and Materials Technology
Standards, Measurements & Testing
Environment
Marine Science & Technology
Biotechnology
Agriculture and Agroindustry
Biomedicine and Health
STD3 - Cooperation with Developing Countries
Non-Nuclear Energy
Security in Nuclear Fission
Controlled Thermonuclear Fusion
Human Capital and Mobility

Source: JNICT

- Scientific Programmes for Stability. Aimed mainly at Greece, Turkey and Portugal
- Programme for Aerospace Technology. This programme was initiated by AGARD (Advisory Group for Aerospace and Development).
- Programmes coordinated by the IPEG (Independent European Programme Group), dedicated to strengthening R&D in topics related to defence.
- Other programmes in collaboration with member countries.

These programmes are presently in vigour.

The greatest leap forward by the Portuguese S&T system came about as a result of the country's entry into the EU. From this point forward, Portugal received a large quantity of funds from the European Commission, the only objective being that of modernising all the socio-economic aspects of the country. In addition to these funds for internal use, Portugal saw the way open to other EU scientific research forums, especially in the EU Framework Programmes.

European Union Framework Programmes

The framework for R&D actions was laid out in the Single European Act, and the Framework Programme was established as the basic instrument for the development, planning and coordination of scientific policy within the Union. The Programme includes the basic areas of high-priority research grouped together in specific programmes designed to satisfy the individual and collective needs of the Member States. Resources to be assigned to the programmes were identified, as were the actions through which the programmes will be developed. The importance of the Framework Programme must not be forgotten, not only as a source of financing, which is enormous, but also as a driving-force element of the Portuguese S&T system. It should also be mentioned that the Framework Programmes, serving as elements of influence on the Portuguese agents, especially in lesser developed areas, carry out an important task of inter-territorial cohesion, although this is not their main mission.

Portuguese participation in the Framework Programmes has steadily increased, clearly showing the development being undergone by the S&T system. It is sufficient to point out that financing by the European Commission tripled between the first and the second Framework Programme. Financing received in the first two programmes, and in the programmes in which Portugal has participated in the third, can be seen in Tables 2-15 and 2-16.

Concentrating on the specific results of the Framework Programme III which ended in 1994, perhaps the best way to summarise its impact on the socio-economic environment of Portugal is to examine the percentage of national participation in comparison with total returns (Figure 2-6).

Figure 2-6: Participation of EU Member States in the Financing of the Framework Programme III.

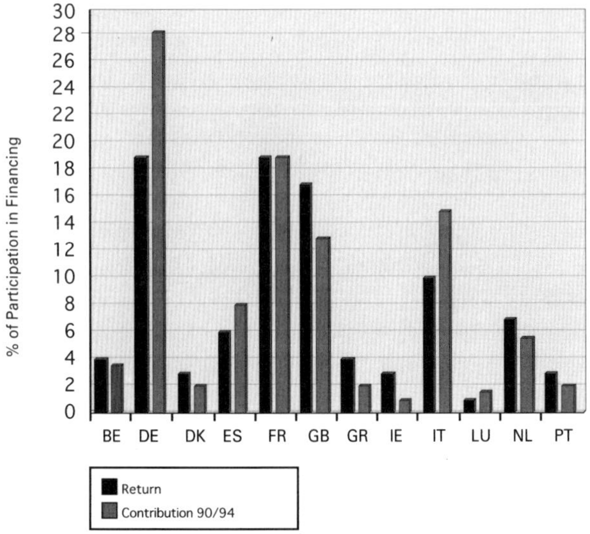

Source: European Commission, 1994

As can be observed, Portuguese participation in spite of not being excessively high when compared to that of countries such as France or Germany, is still significant with 1.4% of the total. This participation has been conditioned by various factors, among which should be highlighted Portugal's relative position with respect to the most advanced countries in this context, the low level of human resources dedicated to this type of programme and the high level of competition reached in the Framework Programmes. The level of returned funds (measured in percentage regarding the total), that is to say, of total financing the quantity returned to the nation in form of contracts, both in the case of public administration and enterprise, reaches 2.4% of EU returns. The difference between financing and returns can initially be qualified as acceptable if compared with the other Member States: in Germany, this relationship is greater than 10%, while in Italy it is 5%. Although investment does not appear excessively high, this must always be measured with respect to the size of the economy. It is not logical therefore to suppose that Portugal should achieve the level of project participation similar to that of the more developed countries, given that overall, it has a lower number of companies, researchers, budget, etc. At the other extreme is France, with a level of returns comparable to that of investment, or the United Kingdom whose level of returns surpasses its contribution. Apart from this series of data, it is fundamental to realise that not all returns are economically quantifiable; a large section may be labelled "intangible" consisting of access to information sources or "know-how" which can be obtained through European partners. Such access comes about by means of cooperation of Portuguese research groups with institutions or companies with high technological or innovation development.

Portuguese participation in the Framework Programme III, with respect to the total of programmes, compared to that of other countries can be seen in Figure 2-7.

Portuguese participation does not even reach 20% of the total of programmes. Comparison with the other countries places Portugal second to last in the ranking, above only Ireland. The reason for such low participation cannot be attributed to the size of the Portuguese economy. Other countries with perhaps lesser developed S&T systems, such as Greece, show higher percentages of participation.

Figure 2-7: Framework Programme III: Participation by Country.

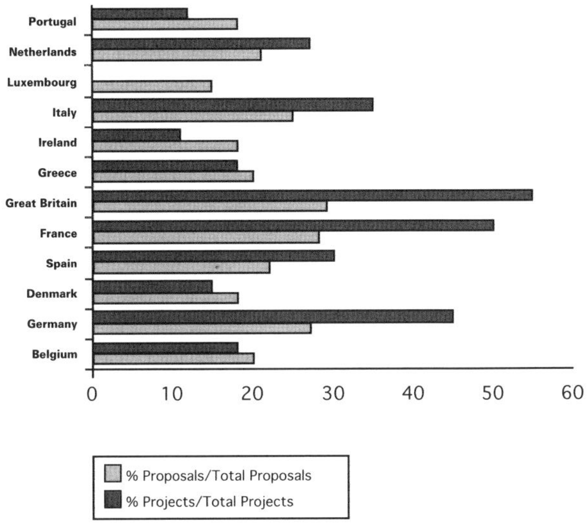

Source: European Commission, 1994

Based on the two graphs, it can be stated that Portuguese participation in the Framework Programmes of the European Union, although acceptable and of quality (return on investment), is still low (percentage of projects with respect to the total). Greater participation in this type of project is necessary at all levels, both public and private, promoting scientific research in the socio-economic environment of the country as a whole.

Portugal's position during the negotiation of the Framework Programme IV and in the present negotiation of V, is basically aimed at the promotion of SME participation in this type of project. Given the nature of companies making up the national productive fabric, the design and implementation of efficient cooperation mechanisms among them, both vertically (within the sector) and horizontally (with other sectors) is essential. The promotion of concrete actions, unconnected to the main EU projects (similar to the CRAFT initiative), is also necessary in order to strengthen this group of companies. Another fundamental aspect to be stressed, is the need for an adequate system of technology transfer to enable the Portuguese S&T system to situate itself at the same level of its European partners.

Other International Programmes

In addition to the Framework Programmes of the European Union, Portugal also participates in other international programmes, and is represented in multilateral organisations and centres. The structure of these programmes does not correspond to previously defined common guidelines, such as in the Framework Programmes, but rather they serve to satisfy the individual R&D needs of countries that are beyond the scope of general EU programmes. The organisational, thematic and administrative structures vary depending on each programme.

The first programme to be highlighted is the EUREKA Project. Initiated in 1985, its objective is that of promoting European competitiveness, facilitating cooperation in R&D among the participating countries. Projects developed under this programme are very close to the final market. In this sense, it is not possible to disassociate actions in this framework from those carried out under EU programmes; they are complementary rather than substitutes. To a certain extent, they cover the actions necessary for the adequate development of S&T systems and which are beyond the range of projects of the European Commission. To date, Portuguese participation in EUREKA can only be considered acceptable. Of the total of more than 900 projects developed so far, Portugal has only participated in 75, with special preference for those related to data processing, robotics, communications, environment, materials, health and biotechnology.

In 1985, Portugal joined CERN. This was a unique opportunity for Portuguese scientists to participate in one of the most up to date institutions in high energy physics world-wide. Since the outset of this participation, and as part of the agreement of adhesion, a fraction of the budget destined to finance the Centre has been used to improve Portuguese infrastructure in areas of activity related to projects developed by the institution. The principal projects in which Portugal has been involved are related to high energy physics, electronics, applied optics, data processing, etc.

Like Spain, Portugal participates in a large number of bilateral cooperation agreements with Latin American countries. Its participation in *CYTED* should be underlined. *CYTED* is a programme which promotes cooperation in the area of applied research and technological development to obtain scientific and technology results which may be transferred to the production systems and social policies of Latin American countries. The programme has three areas of action: thematic networks, pre-competitive research projects and innovation projects.

Portuguese participation in the programme is high, taking part in 30 thematic networks and 17 research projects, involving a total of 94 companies and research centres. As will be observed, participation is highly significant, keeping in mind that 22 countries are involved. The activities developed as a whole mobilise resources to the value of 70 million dollars and the annual budget is 3.5 million dollars.

Table 2-17 shows the participation of the different countries in *CYTED*.

Table 2-17: Participation by country - *CYTED* Programme (1994)

	Thematic Networks	Research Projects	Innovation Projects	Total Activities
Argentina	34	27	7	68
Bolivia	13	3	1	17
Brazil	37	22	2	61
Colombia	29	18	2	49
Costa Rica	23	13	-	26
Cuba	28	12	6	46
Chile	30	26	3	59
Dominican Republic	6	3	1	10
Ecuador	18	8	2	28
El Salvador	10	3	-	13
Guatemala	15	4	1	20
Honduras	11	3	-	14
Mexico	32	25	4	61
Nicaragua	8	3	-	11
Panama	15	5	1	21
Paraguay	13	2	-	15
Peru	20	12	-	32
Portugal	30	17	1	48
Spain	35	28	34	97
Uruguay	17	11	8	36
Venezuela	35	19	2	56

Source: SGPN - Spain, 1994

As in the case of other international programmes, one of the objectives of *CYTED* is the inter-territorial cohesion of the different countries involved in scientific research themes. The programme has brought about a clear tightening of north-south cooperation bonds, making clear its viability and the mutual benefits that may be obtained. The Programme, through its General Assembly, has resulted in the creation of a unique forum for the debate of Latin American R&D policies.

The participation of the *IICT* in cooperation projects with South America should be underlined. In the wake of its international recognition, it now forms part of ECART (European Consortium for Agricultural Research in the Tropics) along with similar institutions in other EU Member States; the United Kingdom, France and the Netherlands. The creation of this consortium took place in Amsterdam on 15 May 1992.

S&T INDICATORS IN PORTUGAL

Analysis of the indicators of the S&T system in Portugal cannot be carried out independently of the recent evolution of the national socio-economic background. In this section, the relationship between the most important parameters showing the evolution of scientific research with those showing the development of the Portuguese economy will be made. Comparison will also be made of the main figures presented with those of other countries in order to evaluate the Portuguese effort regarding R&D.

Portugal is making an important attempt to converge with the other countries of the EU in areas of S&T. Since its adhesion to the EU and in the face of increasingly competitive markets, Portugal has seen itself with the need to make considerable effort in the modernisation of its productive structure and its economy in general. Given that much has already been achieved, the gaps separating Portugal from its EU partners have been reduced. But in spite of everything, Portugal is still at the "tail end" with respect to the other Member States.

The first indicator to be analysed is expenditure in activities related with R&D.

Table 2-18: R&D Expenditure in the period 1982-1992

	Millions of Escudos (Current)	Millions of Escudos (Constant - 1985)
1982	6,541.2	12,295.5
1984	11,307.6	13,773.0
1986	19,867.6	16,487.6
1988	29,910.8	20,007.2
1990	52,032.2	26,945.7
1992	81,201.4	45,847.6

Source: OECD, 1995

As can be seen in Table 2-18, investment practically quadrupled in a period of ten years, in the 80's to be exact, a decade in which greater growth can be observed in many of the OECD countries. From the 90's onward, however, it was Spain which experienced higher growth. Again the differences separating Portugal from its EU partners are great. It should not be forgotten that in international terms, investment carried out is not comparable, neither with larger or smaller countries. Investment in Portugal is a third that of Denmark, or a quarter that of Austria, Finland or Norway.

The differences are somewhat reduced if comparison is made of the percentage with respect to GDP of R&D expenditure. The evolution is shown in Table 2-19.

Table 2-19: R&D Expenditure over GDP (%)

Country	1989	1990	1991	1992
Germany	2.88	2.81	2.83	2.81
France	2.34	2.40	2.42	2.45
EU (Average)	2	2.01	2.1	2.3
Spain	0.75	0.81	0.87	0.92
Italy	1.24	1.35	1.35	1.4
Portugal	0.5	0.6	0.65	0.7

Source: INE - Spain, OECD

The table shows the effort being made by countries in the area of scientific research. As can be seen, Portugal is still far from converging with its EU partners (in more than one percent with respect to the EU average).

One of the best ways of appreciating at a single glance the effort in the area of scientific research is to present in graphic form the trend of the increase in R&D spending in comparison with domestic net growth through the GDP.

Figure 2-8: Breakdown of the Recent Evolution of R&D Expenditure

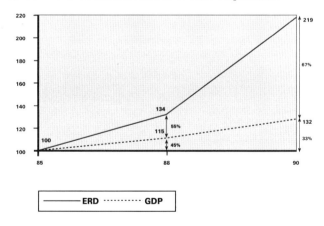

Source: OECD, 1995

The objective of the figure is to show whether GDP or ERD (in percent) undergoes greater growth. Taking 1985 as base with a value of 100, common to both GDP and ERD, in 1987 the latter reached a value of 134 while the former had a value of only 115. In 1990, the difference was even greater with ERD at 219 and

GDP at 132. Analysis of the figure shows that the growth in investment in scientific areas can be divided in two; that inherent to economic growth and that derived from greater effort in the strengthening of investment in research. In 1988, growth in ERD was due in 46% to the growth in GDP and in 55% to greater investment in research. The situation further improved in 1991, when only 33% may be put down to the GDP and 67% to increased investment. In other words, the institutional drive goes even further than pure economic growth.

The general or "structural" tendency of the most important S&T indicators have been outlined so far. In spite of everything, in the last few years (1993-1995), just after the world-wide economic crisis, R&D investment suffered a slight setback. This situation should be considered a consequence and not long term, given that it has come about as a result of the situation and not to a fall off in investment. Forecasts from the Ministry for Science and Technology are very optimistic, foreseeing an increase in investment for 1996 and 1997 (measured in constant prices). This negative data and the perspectives for growth are reflected in the Survey on the Scientific and Technological Potential, now concluded, and which was published in January 1997. Research activities were analysed, at investment level, in 226 companies and 786 research units. According to the survey, the rate of growth of total R&D spending increased 17.59% in 1990 and 9.19% in 1992. On the other hand, in 1995 the falloff was estimated at -0.96%. This drop in the level of investment had greater impact in large companies and public higher institutes.

Figure 2-9: Forecast of the Evolution of R&D Expenditure in Portugal (1992-1997). (Billions of Escudos)

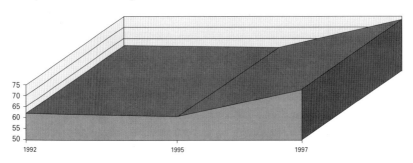

Source: Portuguese Ministry of Science & Technology, 1997
Figures in billions of constant escudos (1990)

Figures in billions of constant escudos (1990)

Source: Portuguese Ministry of Science & Technology, 1997

This increase in effort may seem optimistic, but there is still a lot to be done. An indicator which more accurately measures the efforts of society regarding scientific research would be the percentage of GDP dedicated to R&D, already examined here. Therefore, in spite of experiencing growth, Portugal continues in a clearly inferior position with respect to the developed countries.

The reason explaining these differences is that the proportion of investment is not the same in Portugal as in other industrialised countries with respect to the origin of funds (foreign, public, private). This fact is reflected in Table 2-20.

Table 2-20: Scientific Research Funding.

Country	GDP (M $)	ERD (% of the GDP)	Industry Financed	Government Financed
USA	149,225.0	2.77	50.6	47.1
Japan	62,865.0	2.88	77.9	16.1
Germany	31,585.3	2.73	62.0	35.1
France	23,768.4	2.42	43.5	48.3
United Kingdom	20,178.3	2.22	49.5	35.8
Italy	11,964.3	1.30	47.3	51.5
Greece	336.3	0.47	19.4	68.9
Portugal	501.8	0.61	27.0	61.8
Spain	3,888.8	0.85	47.4	45.1
Belgium	2,751.5	1.69	70.4	27.6
Austria	1,796.7	1.40	53.2	44.3
Finland	1,541.8	1.87	62.2	35.5

Source: EAS, 1991

Table 2-20 shows the imbalance which exists in the origin of research funds. Except for the case of Greece and Portugal, the remaining countries present values of around 50% or higher with respect to the percentage of financing by the productive sector. In Portugal, this value is only 27%, demonstrating scarce interest by enterprise in developing their own technology. As indicated previously, the problem is not that Portuguese enterprise is technologically underdeveloped, but rather, that to date they prefer to acquire from abroad the technology necessary to maintain their international competitiveness.

Yet another imbalance shown by the Portuguese S&T system refers to the distribution of invested funds according to the type of research carried out. Relevant data and its comparison with other countries may be seen in Table 2-21.

The percentage dedicated to applied research is frankly superior to that of other countries with more developed S&T systems. This is due to the higher concentration of resources used in experimental development, of greater use to the national production system. Recent tendencies underline that this unbalanced situation has not been corrected, but rather that the differences have gradually increased.

Table 2-21: **Percentage Breakdown of Research Funds According to Project Type.**

Type of research	Portugal	Japan	Belgium	Spain
Basic research	20.6	14.2	15.1	17.9
Applied research	40.5	27.6	30.2	36.8
Experimental development	38.9	58.2	54.7	45.3
Total	100.0	100.0	100.0	100.0

Source : OECD, 1995

The situation of Portugal also differs with regard to the application of funds taking into account their origin. A type of sealed area has been created in the country, within which "own consumption" of R&D funds takes place; for example, the main part of funds dedicated to scientific research in public administration stays in the public research centres. Likewise, most of the funds dedicated to R&D by business are used in the technological development of the company.

Table 2-22: **Origin and Application of R&D Destined Funds in Portugal (1990)** (Millions of Escudos)

Origin of funds	Application of funds					
	Government	University	PNP	Enterprise	Total	%
Government	12,661.8	17,738.6	863.4	882.1	32,145.9	61.8
University	0.0	321.8	0.0	6.0	327.8	0.6
PNP	33.8	1,253.4	2,839.9	50.8	3,077.9	5.9
Enterprise	181.6	131.5	1,689.2	12,064.8	14,067.1	27.0
Foreign	363.0	402.7	1,065.9	581.9	2,413.5	4.7
Total	13,240.2	18,748.0	6,458.4	13,585.6	52,032.2	100.0

PNP: Private Non-Profit Institutions

Source : OECD, 1995

Public administration finances more than 95% of its own research, while a private company uses 88.8% of capital invested in its own laboratories. The only exception is in private non-profit institutions whose funds are more uniformly distributed. This situation constitutes a serious problem. It is evident that public administration must be able to invest sufficient capital in enterprise to encourage certain areas of research, not only to orient scientific research toward areas of greater national interest, but as a driving-force for future investment by enterprise. It should also be added that certain projects, which due to their nature bring together the different actors through "horizontal" or multi-sectorial programmes, must be coordinated and promoted by public administration. Otherwise, it is almost impossible that they will be carried out. This is especially significant in the case of SMEs, given that without external collaboration, they would be unable to undertake wide research projects alone. (Table 2-22)

The regional distribution of funds is concentrated around the main urban nuclei. The metropolitan area of Lisbon receives more than 65%, while less favoured regions, such as Alentejo, receive only 2.1%. Due to the university sector and a part of enterprise (due to a more satisfactory geographical distribution of activities in the north of Portugal), even greater concentration around the metropolitan areas is avoided.

Table 2-23: Regional Distribution of R&D (1990) (%)

Region	Population	% of R&D Investment	Researchers	R&D Personnel
Lisbon	33.6	63.5	60.8	65.0
North	35.0	18.5	19.4	18.5
Centre	17.5	14.5	15.7	12.0
Alentejo	5.5	1.6	2.3	2.1
Algarve	3.4	0.4	0.7	0.4
Azores	2.4	1.1	0.9	1.6
Madeira	2.6	0.4	0.2	0.4
Total	100.0	100.0	100.0	100.0

Source : OECD, 1994

The number of personnel dedicated to R&D related topics has increased in line with the other national S&T indicators in the last decade. The most significant fact is perhaps the increase in the number of researchers regarding the overall total of personnel. At the beginning of the 90's, the number of scientists increased by 96%, while the overall total of R&D personnel experienced an increase of only 40%. The highest concentration of researchers is in universities, with more than 60% of the total. The tendency toward the concentration of human research capital in the area of higher education has been increasing in recent years. Suffice to say that in 1982, only 42% of total personnel carried out their activities in this sector.

As already noted, R&D investment slightly decreased in the period 1993-1995 due to the background situation already described. Contrary to what may be supposed logical, the forecasts carried out by the Ministry for Science and Technology do not show a setback in the number of researchers, neither in the public sector nor in enterprise. As shown in Table 2-24, the number of researchers in Portugal was 12,042 in 1990. The figure forecast for 1995 was 18,764 and 21,000 in 1997. This means an average annual increase of 10%. (Figure 2-10).

SCIENCE AND TECHNOLOGY IN SOUTHERN EUROPE

Table 2-24: Distribution of Personnel Related to Research (1990)

Occupation	Public		University Administration		PNP		Enterprise		Total	
Scientists	1,094.8	25.9	3,754.6	77.6	622.3	63.8	436.6	21.9	5,908.3	49.1
Technicians	1,835.0	43.4	612.2	12.6	182.4	18.7	1,125.0	56.3	3,754.6	31.2
Others	1,300.1	30.7	473.3	9.8	171.3	17.5	435.0	21.8	2,379.7	19.7
Total	4,229.9	100.0	4,840.1	100.0	976.0	100.0	1,996.6	100.0	12,042.6	100.0
%		35.1		40.2		8.1		16.6		100.0

Source : OECD, 1995

Figure 2-10: Forecast of the evolution of the number of researchers in the period 1992-1997

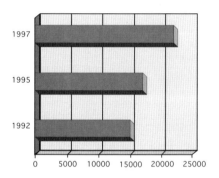

Data for 1995 & 1997 are estimates.

Source: Ministry for Science & Technology, 1997

REFERENCES

Building Bridges .A European Social Democratic Framework for Science and Technology policy. Conclusions from the meeting of EU Ministers for Education

Communication from the Ministry for Science and Technology - 27 January 1997

Decree Law 145/96 Higher Council for Science and Technology

Decree Law 146/96 on the regulation of MCT activities

Decree Law 99/94 on administration of aid within Community Support Framework

Despacho 9/mct/96 on S&T activities in Portugal

Economist Intelligence Unit, Country Profile: Portugal. 1995-1996.

European Commission (1994), Bases del Conocimiento y de la Innovación. COM (94) 378. Communication from the European Commission

European Commission (1995), Green Paper on Innovation.

Eurostat, Europe in Figures (4° Ed.)

Fölster S. (1991), The Art of Encouraging Invention: A New Approach to Government Innovation Policy. The Industrial Institute for Economic and Social Research.

Frascati (1993), Proposed Standard Practice for Surveys of Research and Experimental Development. OECD

Grossman, Gene and Shapiro, Carl (1987), Dynamic R&D, Competition Economic Journal 97.

Information made available by the Portuguese Ministry for Science and Technology in the form of leaflets, magazines, articles, etc.

International Monetary Fund (1996), Perspectives of the World Economy

OECD (1986), Reviews of National Science and Technology Policy: Portugal: 1986.

OECD (1992), Technology and the Economy: Key Relationships

OECD (1993), Proposed Guidelines for Collecting and Interpreting Technological Innovation Data. (Paris)

OECD (1993), Reviews of National Science and Technology Policy: Portugal: 1993.

OECD (1994), Science and Technology Policy: Review and Outlook

OECD (1995), Statistiques de Base de la Science et de la Technologie

OECD, OECD Economic Studies : Portugal, 1996.

Office for Official Publications of the European Communities (1995), General Report on the Activity of the European Union.

Oro, Luis A. & Sebastián, Jesús, Los Sistemas de Ciencia y Tecnología en Iberoamérica. Colección Impactos.

Programme of the XIII Constitutional Portuguese Government

WWW pages of the principal Portuguese public research centres.

GREECE

Chapter 3

GREECE

INTRODUCTION

Of the countries dealt with in this book, and in general in the European Union, Greece is the least developed, and therefore its level of investment in scientific research is the lowest. Nevertheless, the effort made regarding R&D in Greece should not be considered insignificant, neither in a world context nor in relation to the immediate neighbouring countries. The Greek science and technology system as we understand it today, is the most recent of the European Union. Establishing the exact date of initiation would be difficult, given that this calls for the concept to be defined, but in the 1980's a series of events took place in Greece leading to the introduction of what could be called scientific policy. This means that the authorities were made aware of the importance of broadening the scientific and technological capabilities of the country, particularly in such vital aspects as:

- Advanced technologies
- Cooperation between the scientific community and the production system
- Technology transfer.

The first public institutions to regulate this new system were set up in this period, providing the system with appropriate planning and coordination as well as steering the activities of public research centres toward tasks of greater national interest.

The problem to be overcome by the public authorities was the shortage of resources available to develop such activities at the level at which they were being developed in Europe. Faced with this situation, the European Commission through the European Structural Funds (ESF) provided specific financing by means of two programmes, EPET and STRIDE; these were later amalgamated into the programme known as EPET II. The objectives of these programmes were exactly the same as those laid out above, but in inverse order. That is to say, to endow the country with the infrastructure and technology necessary to adapt its competitiveness to that of EU countries, so that at a later date the system could transform itself into an authentic driving force of scientific research.

Much has been achieved in recent years and the incorporation of Greece to common research lines has been effective. However, an underlying problem limiting the development of the Greek system of science and technology still exists. Unlike Spain and Portugal, EU members since 1986 and undergoing rapid economic convergence with the EU average, Greece is diverging in many aspects. Evidently it is going to be difficult to obtain the implantation of a competitive science and technology system. In short, although the system is effective and provides results better than those which would correspond to its levels of investment (one need only consider Greek participation in the R&D Framework Programme in comparison with its financial contribution), it will always be limited unless the country achieves a higher degree of convergence.

Heavy dependency on the ESF has led to the role of the public sector weighing heavily on the system as a whole. This situation can also be seen in other countries although to a lesser extent. At present, measures are being applied in order to decentralise scientific research, currently centred on large urban centres.

The conclusion may be drawn that the foundations for the future development of the system have been laid down but financial structures must be adjusted to the canons established by the more developed countries. Greater development of the Greek economy will bring about a more effective, quality driving force of scientific research.

STRUCTURE OF THE SCIENCE AND TECHNOLOGY SYSTEM IN GREECE

As in the chapters relating to the other countries of Southern Europe, this section outlines the principal institutions that make up the Greek system of science and technology. The flowchart below shows the main agents in addition to the basic lines of communication and coordination. A glance at this chart enables one to see the origin of the different R&D programmes and organisations or institutions responsible for their execution. The principal coordination activities are indicated by the heavy line.

Figure 3-1: Structure of the Greek S&T System

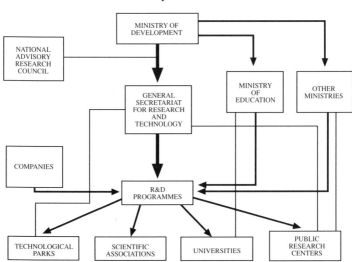

INSTITUTIONAL FRAMEWORK OF THE SCIENCE AND TECHNOLOGY SYSTEM IN GREECE

The institutional structure of the Greek S&T system can be seen in figure 3-2.

Figure 3-2: Principal Institutions Responsible for R&D Policy

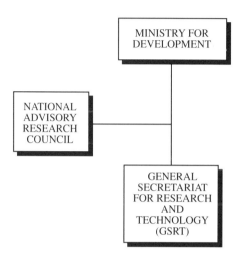

Source: GSRT

Ministry for Development

The Ministry for Development is the main organisation responsible for the promotion, planning and coordination of the S&T system in Greece. Its activities are carried out in coordination with the other Ministries also related in someway to research such as: Energy, Industry and Technology; Economy; Defence; Agriculture; Health; and Culture. It defines scientific policy to be implemented in the country and sets up the most suitable mechanisms for its development. It also coordinates Greek participation in international, bilateral and multilateral projects, both within and beyond the European Union; incorporating such projects into the structure of national scientific projects.

The Ministry for Development is responsible for assigning economic resources to the development of scientific research as well as managing private funds intended for the programmes.

Other missions are that of directing the researcher training policy at all levels, proposing measures for the promotion of employment and facilitating mobility between the areas of research and production.

Annual evaluation reports on the results of the different initiatives carried out are presented to government and parliament. Such reports also contain recommendations regarding proposals and modifications to be introduced.

In order to carry out its activities, the Ministry for Development set up two organisations to provide the support necessary for the correct working of the S&T system in line with national interests and in coordination with activities within the framework of the European Union.

These institutions are:

- National Advisory Committee on Research
- General Secretariat for Research and Technology

National Advisory Research Council

The mission of this organisation is that of promoting the participation of the scientific community, economic and social agents in the elaboration, follow-up and evaluation of national R&D plans. This body is responsible for establishing links between programmed scientific activity and social interests and needs. Its functions include that of making proposals and suggesting objectives to be incorporated into the national R&D plans. It also informs on the fulfilment of these regarding their social and economic impact. At present it is presided over by the Minister for Development and is made up of scientists, members of entrepreneurial and scientific associations, and trade union representatives.

General Secretariat for Research and Technology

The General Secretariat for Research and Technology is the principal Greek organisation in the promotion, planning, and coordination of scientific research. It is a division of the Ministry of Development responsible for the coordination of endeavours aimed at technological development and synchronisation, both in line with the evolution of the Greek economy and the process of European integration. The activities of this Secretariat may be summarised in the following four points:

- Planning and development of national policy regarding scientific and technological research through the design and implementation of national programmes.
- Creation of new infrastructures for activities related to scientific development
- Layout of the general guidelines for national scientific research.
- Technological development of the national production sector.

In addition to these general functions, the GSRT is entrusted with programme funding. It also acts as an interface between the scientific community, enterprise involved in public funded projects and public administration. Programmes aimed at main economic sectors with greater potential are considered to be of high-priority as regards funding.

Other objectives such as technology transfer and agreements among companies to jointly develop research have acquired greater importance, as has the elaboration of bilateral or multilateral agreements in scientific cooperation with other countries.

As in the case of other national organisations responsible for the promotion of R&D, the GSRT also disseminates scientific and innovative advances in order to foment an "authentic technological culture". It creates, supervises and finances public research centres as well as technology parks specialising in research and development. Enterprises were set up, "Technological Companies for Technological Research and Development", entrusted with providing technical assistance to SMEs (the great majority of Greek companies) and solutions to the specific needs of the production sector. Four technology parks have been created to endow the national framework of specific research centres with a high technological level enabling them to develop lines of research which otherwise they would be unable to do depending solely on their own resources. Not only have companies been provided with a series of aids or effective channels of technology transfer, but a suitable mechanism has also been established to enable them to develop the results of scientific research commercially.

The organisational structure of the General Secretariat for Research and Technology can be seen in Figure 3-3.

Figure 3-3: Structure of the General Secretariat for Research and Technology

```
                    ┌──────────────────┐
                    │     GENERAL      │
                    │ SECRETARIAT FOR  │
                    │   RESEARCH AND   │
                    │    TECHNOLOGY    │
                    └────────┬─────────┘
                             │
┌────────────┬──────────────┬──────────────┬────────────┬──────────────────┐
│  PLANNING  │    R&TD      │ SUPERVISION  │            │  TECHNOLOGICAL   │
│DIRECTORATE │ PROGRAMMES   │   OF R&TD    │ TECHNICAL  │   DEVELOPMENT    │
│            │ DIRECTORATE  │ INSTITUTIONS │  SERVICE   │   DIRECTORATE    │
│            │              │ DIRECTORATE  │            │                  │
└────────────┴──────────────┴──────────────┴────────────┴──────────────────┘
┌────────────┬──────────────┬──────────────┬────────────┬──────────────────┐
│INTERNATIONAL│             │ DIVISION OF  │            │    COMPUTER      │
│    S&T     │ADMINISTRATION│ PUBLICATIONS │  DIVISION  │      AND         │
│COOPERATION │ DIRECTORATE  │ CONFERENCES  │ STRUCTURAL │  ORGANIZATION    │
│DIRECTORATE │              │     AND      │ PROGRAMMES │  DIRECTORATE     │
│            │              │ EXHIBITIONS  │            │                  │
└────────────┴──────────────┴──────────────┴────────────┴──────────────────┘
```

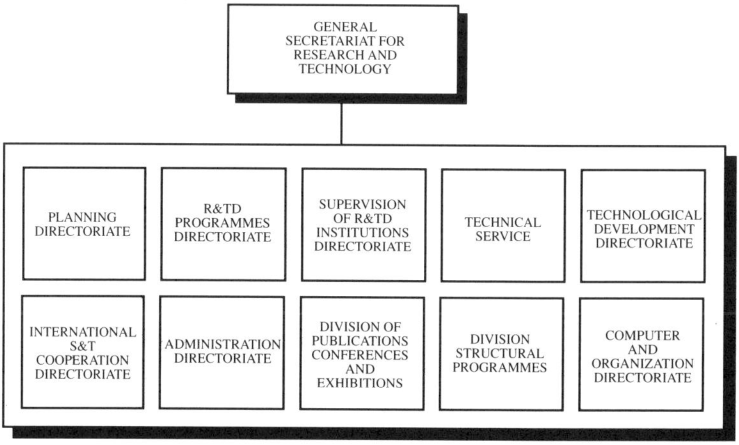

Source: GSRT

THE PUBLIC SCIENCE AND TECHNOLOGY SYSTEM: PUBLIC RESEARCH CENTRES AND MAIN R&D PROGRAMMES

In the last decade one of the principal problems attributed to the Greek S&T system was the lack of adequate integration and coordination of public activities in the area of scientific research. Public research centres were usually independent bodies following their own initiatives and guidelines and lacking adequate coordination in the elaboration of projects which took into account national socio-economic impact.

Only at the end of the 80's and beginning of the 90's did the Greek government begin to gradually reorganise public research at the suggestion of the European Union. The first step toward global coordination of the system was the creation of the post of Minister for Development. Specific organisations, similar to the General Secretariat for Research and Technology, were set up to control, supervise and plan R&D. It can now be stated that Greece has established the minimum requirements to enable it to develop an authentic S&T system that could provide the country with the conditions necessary to be competitive in the framework of the countries of its area, especially within the European Union. Greek public administration has laid down the basis for future growth in the awareness that the need to promote innovative and technological change is a basic need of all economic systems.

Below is a description of the principal public research centres in Greece and their main R&D activities. The mobilising impact on resources that this type of institute has on the socio-economic environment, especially in those sectors directly related to the business or production sector, should not be forgotten when analysing the work of publicly financed foundations or institutes.

Public Research Centres

The main public research centres in Greece come under the supervision of the GSRT. The most important are listed in Table 3-1.

Table 3-1 Principal Public Research & Technology Centres

National Centre for Scientific Research Demokritos
National Observatory of Athens
National Hellenic Research Foundation (NHRF)
Foundation for Research and Technology (FORTH)
Institute for Language & Speech Processing (ILSP)
Hellenic Pasteur Institute
National Centre for Social Research
Alexander Fleming Foundation for Basic Biology
Crete Institute for Marine Biology (IMBC)
National Centre for Maritime Research (NCMR)
Institute for Chemical Engineering and High Temperature Processes
Institute for Chemical Engineering Research

Source: GSRT

The activities of the principal research centres are outlined below. Due to the large number of such centres, detailed descriptions have been limited to the most important.

National Centre for Scientific Research - Demokritos

The Demokritos Institute was created in 1958 as a branch of the Atomic Energy Commission of Greece. Its present structure came about in 1988 as the result of a parliamentary act. Although the Institute depends on the General Secretariat for Research and Technology, it operates independently and autonomously under the supervision of the Ministry for Industry, Energy and Technology.

Its activities fall within the following areas of science:

- Biology
- Data Processing and Telecommunications
- Materials Science
- Microelectronics
- Nuclear Physics
- Nuclear Technology and Radiation Protection
- Chemistry
- Radioisotopes and Radiodiagnostic Products

Activities are carried out in 8 independent institutes.

Overall management is in the hands of a Board of Management made up of the Directors of the various institutes. Of the 700 staff employed by the Centre, 180 are full-time researchers. Staff directly related to scientific research is 325, almost half of the total.

Initially the objective of the Centre was the solution of concrete problems of Greek industry although in recent times it has undergone a gradual opening up process toward foreign cooperation. At present it collaborates with the International Atomic Energy Agency, thanks to which it has also signed numerous cooperation agreements, especially in the framework of the European Union. Attention should also be drawn to its high level of participation in European R&D programmes such as BRITE, EURAM, ESPRIT, ECLAIR, etc. and which in one way or another have marked the areas of research in which it works.

The areas encompass a wide spectrum of science with specific projects covering areas going from basic research to research aimed at final products.

National Observatory of Athens

The National Observatory of Athens is a public institute carrying out research in areas related to astrophysics. It is directed by university teachers, the majority of whom are from the Technical University of Athens.

The National Observatory of Athens consists of four independent institutes, each one with its own installations.

- The Astronomy Institute studies the stars and weather.
- The Meteorology Institute carries out studies explaining climatological behaviour and related atmospheric phenomena.
- The Geodynamic Institute researches seismology and geophysical phenomena.
- The Institute of the Ionosphere carries out systematic studies on the different layers of the atmosphere.

At present more than 100 researchers are employed by the Observatory.

Foundation for Research and Technology

The Crete Centre for Research began its activities at the end of 1983 within the activities of the University of Crete. It initially consisted of three institutes and had as its main objective the strengthening of university research activities as well as to serve as a catalyst for the development of Crete and the country in general. In 1984 and 1985, four new institutes were added to this centre. In 1987 the Centre changed its name to Foundation for Research and Technology (FORTH).

The headquarters is located in Heraklion, Crete. The 7 institutes that make up FORTH are outlined below.

Table 3-2: Network of FORTH Institutes

ICS:	Institute of Computer Science (Heraklion)	
IESL:	Institute for Electronic Structure and Laser (Heraklion)	
IMBB:	Institute for Molecular Biology and Biotechnology (Heraklion)	
IACM:	Institute for Applied and Computer Mathematics (Heraklion)	
IMS:	Institute of Mediterranean Studies (Rethimnon)	
CPERI	Institute of Chemical Engineering Processes (Thessaloniki)	
ICE:	Institute of Chemical Engineering (Patras)	

Source: FORTH, 1996

The Foundation has acquired justifiable prestige in international science, due mainly to the dissemination of the technological results of research through the different technological parks established in Greece. As will be indicated further on, these constitute an important element in the Greek S&T system, serving as an effective interface between public R&D funding and the sectors of production and enterprise.

Hellenic Pasteur Institute

The Hellenic Pasteur Institute was constituted in 1919 under private initiative and was known as the Greek Institute of Microbiology, supervising the study of infectious diseases and vaccines. It became public in the 40's after serious economic problems. In the 70's an agreement was signed with the Pasteur Institute in Paris, giving rise to the current title. Collaboration with the latter has been decisive in recent times to the extent that part of management is made up of French scientists. The principal research areas are:

- Public Health Care.
- Training and Education Programmes.
- Biomedical Research.

At present, it employs more than 100 scientists.

National Hellenic Research Foundation

The National Hellenic Research Foundation is involved in a wide spectrum of science related activities. As opposed to other institutes or foundations, studies and programmes directly related to humanities are carried out. The NHRF consists of six independent centres:

- Organic Chemistry
- Biological Research
- Byzantine Research
- Studies of Rome and Ancient Greece
- Neo-Greek Research
- Theoretical Chemistry and Physics-Chemistry

Association for Research, Technology & Training

The Association for Research, Technology and Training (ARTT) was created in April of 1991 in Heraklion, Crete, as a non-profit making organisation supporting the S&T system in Greece, its main objectives being as follows:

- To support and develop scientific research in Greece and other EU countries.
- To promote, transfer and disseminate technology products, services and expertise.
- To develop specific training projects for scientists and researchers.
- To support specialised training in universities and research centres.
- To make available high-quality specialised services, such as consultancy, management or technological audit for scientific communities in areas relating to infrastructure, project development or training plans.
- To promote cultural activities and publications in topics directly related to scientific research.

ARTT is affiliated to the Foundation for Research and Technology (FORTH), and has access to their training infrastructure, in addition to the expertise gained by the scientists in carrying out their work. It should be pointed out that a large majority of these scientists were responsible for the creation of this Association. The staff of ARTT has acquired ample experience in technology transfer projects through their involvement in more than 120 EU financed projects. ARTT has participated actively in programmes such as ESPRIT, BRITE, EURAM, SCIENCE, JOULE, THERMIE, SPRINT, COMET, ERASMUS and others. Since 1991, it has provided back up services to companies in training and research based on the experience of its scientists and on the infrastructure available. To date, services provided to public research centres and companies have been concentrated on:

- Viability Studies
- R&D Project Management
- Technical Consultancy
- Project Evaluation
- Design of Prototypes and Technology Products
- Personnel Training in areas related to:

    - Data Processing and Applied Mathematics
    - Microelectronics and Devices
    - Laser Technology and Applications
    - Biotechnology

Although ARTT is located in Crete, activities are developed throughout Greece especially in the regions of Athens, Salonica and Patras where the greater part of enterprise from the production sector is concentrated.

At present the Association is directly responsible for carrying out and managing projects financed by the European Commission and in particular relating to the STRIDE, BRITE /EURAM, FAST, and PRAYS programmes.

It also participates in the following European programmes:

- RADIO - PRAYS
- Information and Organisation of Rural Distribution
- MONITOR FAST
- Cohesion in Europe
- STRIDE-STRIDEL
- HELP. Hellenic Project for a greater application of R&D
- Hellenic Action for the multimedia applications and information systems
- BRITE-EURAM
- OSMOS: Optic Fibre Sensing System for Monitoring of Structures
- HUMAN CAPITAL AND MOBILITY

It is also involved in national studies, the most important of which are:

- Viticulture in Crete
- Development Problems and Perspectives
- HEART
- Horizontal European Activities in Rehabilitation Technologies.

Technology Parks in Greece

One of the assignments of the General Secretariat for Research and Technology was the creation of technology parks where a series of research units, pioneers in the area of R&D, were to be installed. The thinking behind this was that of bringing the area of production closer to the research institutes and thus enable more fluid exchange of expertise and know-how. The existence of these technology parks has permitted more efficient exploitation of the results and products of the different lines of research carried out in Greece.

The three parks presently in operation are listed below:

- Science and Technology Park of Crete (Heraclion)
- Science and Technology Park (Thessaloniki)
- Science and Technology Park (Patras)

Given that the functions of all three are similar, the description of one, that of Crete, will be used as an example.

Crete Technology Park

The idea of creating this park can be traced back to 1988, when certain members of FORTH, one of the most prestigious research institutes of Greece, decided to create a centre that would serve as an interface between public research and the production sector. In order to achieve this objective, the company STEP-C was established in the park in December 1993, with an initial capital of 40 million drachma. Apart from technology transfer, these parks have also served as a driving force for regional development, in this case the island of Crete. STEP-C, responsible for the management of the park, is made up of companies from the Greek production sector. This permits exploitation of the opportunities made available by technological institutes, converting the latter into vehicles for the dissemination of science.

In spite of the fact that the principal role of the park is the dissemination of the knowledge accumulated by the scientific community, transforming it into an authentic "third pole" of scientific research, support programmes have also been developed for local industry especially in the area of distribution and services.

This *third development pole* would be incomplete if only the companies established within, participated in its activities. To avoid this situation, the participation of external companies was proposed, thus bringing about the function of the parks as authentic entrepreneurial associations, both on a sectorial and multi-sectorial basis and as a forum for the exchange of expertise and know-how. R&D is the uniting link.

STEP-C collaborates at present with both private and public organisations. It provides services to 18 companies as associated members. Areas of cooperation are based on the joint development of projects related to the following areas:

- Electronics.
- Medical Equipment.
- System integration
- Software development
- Telematics and Telecommunications
- Applied Mathematics
- Industrial Automation
- R&D Project Management

FORTH also participates in this technological park, and to date has offered the possibility of formalising collaboration contracts between companies and the institutes associated with the Foundation in topics related to:

- Computer Science
- Laser Microelectronics and Applications
- Polymers
- Development of Medical Prototypes
- Biology and Biotechnology
- Applied Mathematics
- Telematics and Computer Networks

Given the success of the initiatives sponsored by this type of park, it is planned to establish a greater quantity, spreading the benefits of participation throughout Greek business.

MAIN PUBLIC R&D PROGRAMMES: THE EPET II PROGRAMME

One of the "traditional evils" of the Greek public science and technology system was the lack of adequate coordination of research related activities. International experience has shown that the best way of approaching planning of such efforts is through specific national plans which identify the most relevant areas of research for the development of society. It should not be forgotten that science and technology have become an essential item of expenditure in any industrialised country. It could even be stated that this is a new sector of activity which energises and impels the economic growth of other sectors. In other words, it functions as a driving force both in development and in its capacity to mobilise new resources toward other areas of interest.

The first attempt to introduce appropriate planning into the Greek science and technology system took place in 1979 through the Agency for Scientific and Technological Research (YEET). This organisation was also responsible for the design of national scientific policy and launched the first phase of the National Science and Technology Programme (EPET) for the period 1970-1980.

The keys of this programme were twofold:

- Establishment for the first time of a concrete policy for science and technology, designed to improve infrastructure and determine priorities in line with government established objectives.
- Implementation of public funding for projects based on the criteria of quality, giving special preference to activities demanding the intervention of various institutes or companies.

Due to various bureaucratic and funding problems, the result of the first initiative was not effective. It was however, the first step toward research planning.

Greece initiated its current series of national R&D plans under suggestions from the European Commission as a result of its adhesion to the EU. These plans were adapted to the general guidelines of other member state programmes.

This implied intensive financing on the part of the European funds as it must be remembered that GDP per capita in Greece is only 52% of the EU average. The mechanism employed for this financing has been the Community Support Framework.

The objective of the Community Support Framework is that of financing investment in human and physical capital in order to promote growth and reduce the economic gap between Greece and the more developed EU countries until total convergence is reached. In order to make this possible, it is not enough that the country develop at same rate as the other countries, but rather growth must be greater and supported.

Modern economic theory emphasises the positive impact of certain factors on the growth process, for example; private investment, quantity and quality of infrastructure, quality and competitiveness of human capital, the efficiency of institutions and an adequate level of development in science and technology.

The National Science and Technology Programmes were redesigned taking into account these principles. The former title, EPET, was kept. At present, Greece has already developed the first phase (EPET I: 1989-1993) and is now involved in the second phase (EPET II: 1994-1999).

The primary objective of this programme is to increase the level of funds allocated to R&D. Before the initiation of EPET II (1994), the quantity invested through the EPET operational programme of science and technology was 120 billion drachmae of the total figure assigned by the ESF to Greece (1.7% of the total provided by the European Commission for this country). This percentage was less than the average gross expenditure of European funds on R&D in other

EU Member States. Since the Community Support Framework consists of a series of pre-budgeted funds invested during a specific time period, it was necessary that the proportion intended for scientific research be increased significantly, especially given that science and technology systems have become crucial mechanisms for sustained growth. It was estimated, furthermore, that the materialisation of adequate investment would be particularly effective and would improve Greek competitiveness, strengthening economic activity as a whole. This would eventually lead to an increase in public and private investment in technology, creating a lead-on effect of economic growth, technological innovation and promotion of R&D.

The present National Research and Technology Programme (EPET II) was designed with this objective in mind. Furthermore, it complies with the budgetary restrictions imposed by public expenditure; 114 billion drachmae plus 6 billion allocated to regional investment.

In addition to the primary objective of increasing investment, another series of facts was taken into account when designing R&D institutional policy. The main ones are outlined below:

- Greece's endeavour in the area of R&D is not sufficient to raise it to the level of other EU countries.
- The new political situation in the Balkans and Middle East has forged a new role for Greece in these areas.

This situation determined the policy to be followed. It is outlined by the following points:

1. To enhance cooperation between different R&D organisations and the production sector in order to develop projects of high economic interest (i.e. environment, new materials, information technologies, telecommunications, biotechnology, etc.).

2. To encourage the transfer of technology from abroad through:
   - Execution of new technology projects with foreign partners
   - Improve international technical assistance, licensing, etc.. .
   - Widen coverage to include projects which imply high economic risk due to the adoption of new technologies.

3. To improve and strengthen the innovative capacity of Greek companies in the development of actions and infrastructure leading to the integration of innovation.
   - To support the innovative process in all its stages through mobilisation, awareness campaigns, technical support, etc.

4. To introduce information and evaluation systems to ensure correct use of public funds:

   - Purpose designed information networks, data bases, publications and conferences with the high-priority objective of research dissemination.
   - Establishment of special units in universities and public research centres for the dissemination of scientific knowledge.
   - Creation of innovation centres, technology parks, agencies specialised in supporting programmes of high economic risk and the provision of know-how and consultancy services to innovative companies.

5. To support and restructure sectorial initiatives through special initiatives and in such a way that the challenges and needs of specific sectors can be broached thus creating competitive advantages (e.g. telecommunications, energy, environment, biotechnology, new materials, administration, social space, culture, sports, etc..)

   - Given the overconcentration of R&D activities in the region of Attica (50-55% of total national resources), an effort should be made to optimise the structure of resource allotment as well as the promotion of regions which may have influence with respect to Greece's position within the framework of the EU, the Balkans and the Mediterranean area.

6. To tackle the training needs of human capital by means of concrete programmes in new technologies and techniques, with special emphasis on post-graduate research carried out by young scientists.

7. To support the cultural assimilation of new communications and information techniques resulting from innovative processes. Special emphasis should be given to the importance of initiative and innovation in the field of education. These objectives should take the form of special actions which encourage technological culture.

   Special effort will be made to coordinate the development of projects, which because of their size, are included in several initiatives of the Operative Programme, e.g.: natural gas, energy distribution channels, etc. The aim is to eliminate redundancy and overlapping project management, a common problem in the past.

   - Particular emphasis will be given to the development of human resources by promoting initiatives in matters relating to research and technology, taking into account the evolution of relations with the Balkan countries and the Middle East.

These guidelines of R&D public sector policy are made reality through the National Research and Technology Programme (EPET II) described below.

This operational programme is split into four chapters, five subprogrammes and various measures and actions.

The first chapter makes reference to the important role played by R&D in the national economy and provides a brief overview of the main indicators of the S&T system. It describes the existing infrastructure with its strengths and weaknesses and shows the institutional structure supporting the system. The second chapter makes reference to the European R&D financing mechanisms used to date (competitive programmes, framework programmes, Structural Funds, community initiatives, etc.) and analyses national policies applied in the past. The principal conclusions regarding the impact of EU programmes on scientific research as a whole in Greece are also laid out. The following chapter shows the strategic objectives sought after by EPET II. After a short analysis of policy to be developed, the compatibility of the objectives of both programmes is established and the manner in which this programme can contribute to its application. The programmes making up EPET II are outlined as well as the evaluation criteria that will be employed. Finally, the fourth chapter contains a detailed description of the five subprogrammes making up the R&D national plan.

Given the scope of this book, description will be limited to the general guidelines of each programme, more so from the point of view of impact and motivation rather than the technical aspects.

Subprogramme 1:
R&D in specific areas

One of the characteristics of R&D public funding in Greece is the high proportion of foreign funds supporting the national programmes. At present this contribution reaches 20% of the total. Given that in 1986 this percentage was only 2%, it can be asserted that growth has been considerable and was in fact the highest of all the EU Member States. The contribution of foreign capital to the maintenance of this type of activities is even more significant in the business sector where it reaches 23%. Such favourable conditions means that Greece has reached a level where it can "offer" the results of research to European companies. For this to occur, deep changes are needed in the structure and organisation of the public research centres in Greece in order to benefit from the sale of patents and services.

The National Research and Technology Programme should therefore make greater effort in selected areas of higher economic profit.

According to studies carried out by the General Secretariat for Research and Technology, R&D needs in Europe are to be found in the secondary sector and not in the tertiary, which incidentally is at a level comparable, or even higher, than that of the United States or Japan. In this context EPET II concentrates its activities on the following specific fields related to research:

- Biological Sciences. In particular, health and agriculture, with special emphasis on biotechnological applications.
- Information technologies, industrial applications and services
- New materials or improvements in existing ones and development of new manufacturing processes.
- Analysis of the impact of the social, administrative, economic and cultural features of development. This area is developing rapidly, due mainly to the adjustment necessary between the emerging innovative processes and assimilation by society, especially by the production sector.

The objective of this first subprogramme is to strengthen activities in the areas just mentioned. However, the problem lies in that some are too general to establish the concentration of resources needed to create competitive advantages. The programme on occasion fails to make clear reference to specific projects, but rather to the general guidelines that have to be followed. This, which initially could be an advantage, implies the risk of research not being totally oriented towards satisfying national needs. In order to avoid such fragmentation of scientific research toward areas of little strategic importance, greater significance has been attributed to the co-financing of these projects by enterprise, in such a way that it is the companies themselves who risk their capital in activities, potentially advantageous on an individual basis.

To date this is the second most important subprogramme, having received 26% of allotted funds and 80 projects approved of the 750 received.

Subprogramme 2:
Industrial Research/Technological transfer/Innovation.

The creation of new innovative processes in the production of goods and services demands an adequate level of technological transfer and dissemination of research results. This transfer must be accomplished not only at the domestic level (among public research centres or universities and enterprise), but also internationally. In order that a S&T system can be efficient, the exchange of knowledge, especially with international linkages, must be balanced in such a way that the demand for new production or organisational processes be equivalent to the contribution by the other actors.

The objectives of this subprogramme are to:

- Create networks of know-how and concrete information channels on R&D
- Encourage the development of data and library bases related to technological aspects of industrial interest.
- Promote the creation of a National Documentation Centre

Another fundamental aspect is the creation of specialised units within industry providing support to companies in concrete topics, such as specialised consultancy services: quality control, reverse engineering, technical documentation, technology management, etc. These services will be provided through research and development agencies created to this end, technology parks, technology transfer parks, quality control and certification laboratories or other related entities such as ELOT, OBI, EOMMEX, ELKEPA, the last being specialised technology centres.

The development of this subprogramme demands careful management due to the high number of companies involved. For this reason, various operational programmes have been created within EPET. These programmes coordinate actions on an independent basis. Special importance has been attributed to initiatives offering integral solutions to specific problems, eliminating superfluous initiatives or intermediary organisations. The latter has been considered fundamental, since in the past, one of the main problems of the S&T system in Greece was the growing demand for new technologies or innovative processes unmatched by supply in spite of the fact that the latter was also growing.

The creation of units or financial entities, offering tailored back-up to companies in order to enable them to participate in R&D projects, will also be supported.

Another R&D policy measure is the creation of an appropriate environment to incorporate new technologies and disseminate existing ones. The creation of new communication links are vital between public research centres and national production. Such links will be promoted through the following initiatives:

PAVE (Industrial Research and Development Programme).

The objective of this programme is the promotion of industrial initiatives on any scale. Initiatives responding to the needs of a specific industrial unit (bottom-up approach) will receive special support. In this sense PAVE, framed within the national plan EPET II, differs substantially from that framed within EPET I. In the latter, the initiatives aimed at industry took the form of large sectorial initiatives whose object was to encourage research in areas of special strategic importance (up-down approach). The reason for the change in approach is to complement the initiatives developed in the first programme and make them more accessible to industry in general.

YPER (Scholarships for Guided Research)

This is a new programme within EPET II. The principal objective is to promote applied research within industry. In order for a project to be accepted under this programme, it must have the participation of a company, a public research centre and a research student, wishing to carry out their doctoral thesis on a topic of interest to all three parties. This promotes links among three of the basic agents of science and technology: university, enterprise and public research centres. In addition it facilitates the incorporation of young scientists into research.

SYN (Co-financed Programmes)

One of the basic aspects of the development of large research projects of interest to an entrepreneurial sector is cooperation among companies of that sector and public research centres. The objective of the Co-financed Programmes is to stimulate a system which enables this type of initiative to be carried out through mixed public-private financing. Given that SYN continues the initiatives set up in the first phase of EPET, a special effort has been made to increase coordination of actors participating in the development of new products or innovative processes. The basic objective is to avoid repetition and to ensure that pubic funding acts as a true economic driving force, not simply a non-risk means for companies to finance R&D.

Liaison Offices

As mentioned above, the relationship between research developed in public centres, universities and the production sector should be strongly linked to specific needs of national enterprise. Liaison Offices have been set up under EPET II to improve bridges of communication and cooperation between public and private entities. The industrial application of the results of public financed research in government centres is promoted. Special Information Centres have also been established to complement the activities of the Liaison Offices. These are government organisations or state run enterprises which promote the best possible application of the results of state research contracts.

As has already been noted, economic development does not depend solely on the mobilisation of resources by enterprise and the introduction of new techniques or products. It is necessary to create an authentic business technology culture and make public opinion aware of the importance of scientific research. This is still at quite a low level in Greece compared to neighbouring countries. It is in the context of this subprogramme that future initiatives will be implemented in order to reduce the gravity of the situation.

At this point, attention should be drawn to the fact that the basic purpose of this second subprogramme, apart from the promotion of R&D in the production sector, is the development of structures for technology transfer and the creation of efficient communication links, which did not exist in sufficient number between the different actors developing scientific research.

Subprogramme 3:
Support and Restructuring of Research Infrastructure

As in the case of all the lesser developed Member States, the European Union has provided new sources of funding for R&D activities. In particular, Greece through this new funding has developed its research infrastructure in the last decade, as well as reorienting university activities toward this field.

The purpose of Subprogramme 3 is to move a step forward in the amplification of structures necessary to develop science. The actions therefore proposed are the reinforcement of communication through coordination and cooperation of public research centres, guidance and specialisation of organisations involved in research in areas of strategic national importance and high economic interest, decentralisation and regional development of scientific infrastructure, harmonisation of activities with the productive needs of the country, and finally, the development of human capital in research.

More specifically, Subprogramme 3 includes the following:

- Reorientation, restructuring, and eventual broadening of the existing R&D infrastructure, aiming at the location of R&D structures in line with the economic interest of activities being developed in the zone. This development will be based on the analysis carried out by the state on the levels of investment incurred to date in the research and technology agencies, their perspectives, contribution to the upgrading of the national S&T system, the international scope of their activities and the procedure for the systematic evaluation of the activity developed.

- Establishment of new R&D organisations, complementary to those already existing, both at a functional and regional level, and avoidance of the creation of research centres or organisations of little use to national interest, as has been the case in the past. The need for exhaustive studies to be carried out previously demonstrates the general interest in these new organisations.

Another of the objectives of this subprogramme is that of the decentralisation of R&D. At present the majority of scientific activities are carried out in the urban nucleus of Athens. This fact, which is significant, is not unusual given that the concentration of this type of activities in connection with the large urban nucleus is a world wide phenomenon. In spite of everything, the Greek Government is endeavouring to bring scientific research to zones where its development could be more beneficial, especially where industry needs it. To be more specific, these are the regions of Epirus, Macedonia and Thraki, as well as certain regions of the south where greater funding will be available to establish new research centres.

The restructuring and decentralisation of the Greek S&T system must be compatible with high-priority national objectives. The position of the country must be assured in the face of the new Balkan countries and those of the Middle East.

Subprogramme 4:
Human Capital.

Owing to the fact that in Greece, the level of R&D infrastructure is not exceptional, as would be the case in the more developed EU countries, the active factor of the Greek system of science and technology is human capital. This fact is corroborated by the type of research carried out in this country which generally is much more intense in human capital than in infrastructure. The continuous improvement of human potential in research is vital for any nation. From a quantitative point of view, it is expected that by 1999, when EPET II concludes, the number of researchers per 1,000 inhabitants will have increased from 2.4 to 3.5.

The question is not only that of increasing the global number of researchers but those involved in areas where their work is of greater use. The overall objective of the Human Capital Programme is to increase the number of scientific personnel employed in companies by 50 to 100%. However, achieving this target number of scientists with adequate training is not sufficient. Enterprise must also be aware of the need to employ them. In other words, this subprogramme has two aims; the first of which is training and the second, making people aware.

Numerous training and qualification programmes have been designed closely adapted to the different needs of each sector. Activities will be as follows:

- Training of new research personnel as well as training and reorientation of the trainers responsible for instructing them on rapidly evolving new technologies.
- Training of specialised technical personnel who can provide support in R&D activities and participate in the application of new technologies.
- Training of experts in new businesses related to technology, especially in sectors of greater interest to the sectors of industry and services in Greece.
- Training and qualification in innovation and R&D management.
- Establishment of exchange networks of staff in laboratories whose activities are in the same field.
- Encouragement of mobility of research personnel among public research centres and industrial sectors making up the industrial fabric in Greece.
- Promotion of concrete actions facilitating the mobility of Greek research personnel abroad and that of foreign researchers in Greece with special emphasis on those from Central or Eastern Europe. Reference is also made to facilities to be provided to foreign scientists who can provide know-how on the development of state of the art technology.
- Creation of mechanisms to evaluate proposed training initiatives, as well as the carrying out of comparative studies among the staff of similar research centres.

The global objective is to equip the Greek S&T system with a broad training programme for scientists and research personnel. Previously, no one programme of such magnitude existed: the budget intended for this activity is almost 10% of the total of EPET II. Given its pilot nature, future programmes will be needed to define actions and orient them.

The estimated budget of the sub-programmes of EPET II, the main source of public funding in R&D, is outlined in Table 3-3. The total is 580 Mecus, of which it is estimated that 27% is from the entrepreneurial sector in the form of participation agreements. The remainder originates from public funds especially those of the European Union.

Table 3-3: Community Support Framework for R&D (EPET II) *

	Billions of drachmae		Billions of ECUs	
	Public 122,35	Total 168,93	Public 421,879	Total 579,068
Subprogramme 1: **R&D in specific areas**	**35,99**	**43,63**	**124,113**	**150,441**
Environment & Quality of Life	7,20	8,73	24,823	30,088
Life Sciences	6,17	7,48	21,277	25,90
Information Technologies	10,28	12,47	35,461	42,983
New Materials	10,28	12,47	35,461	42,983
Culture, Society & Technology	2,06	2,49	7,092	8,597
Subprogramme 2: **Industrial Research, Technological** **Transfer & Innovation**	**43,85**	**72,92**	**151,223**	**265,245**
Industrial Research PAVE	13,79	27,58	47,553	95,106
Applied research YPER, SYN	3,50	5,44	12,056	16,548
Technology Transfer	22,45	39,74	77,429	137,042
Networks, Data Bases, Documentation	2,06	2,42	7,092	7,092
International Cooperation in R&D	2,06	2,74	7,092	9,456
Subprogramme 3: **Infrastructure Support/** **Extension of R&D**	**26,74**	**19,57**	**92,199**	**101,950**
Support to priority Areas	11,31	14,14	39,007	48,759
Extension in North Axis	10,28	10,28	35,461	35,461
Extension in South Axis	5,14	5,14	17,730	17,730
Subprogramme 4: **Human Capital**	**11,65**	**13.70**	**40,160**	**47,247**
Training of R&D Research Personnel	8,56	10,07	29,522	34,732
Mobility, Links with Production	3,09	3,63	10,638	12,516
Subprogramme 5: **Management of CSF**	**4,11**	**4,11**	**14,184**	**14,184**
Administration Follow-up	2,06	2,06	7,092	7,092
Evaluation, Evaluation, Studies	2,06	2,06	7,092	7,092

Note: *Public spending does not include 6 Billion ECUs assigned at regional level*
Rate of exchange: 1 ECU=290 drachmas.

Source: *Summary of the National Programme EPET II.*

* Drachma. Average exchange rate in 1994: 273 drachma =1$; 158 drachmae =1 Deutschmark; 309 drachmae =1 ECU

OTHER NATIONAL PROGRAMMES

Research consortium for Improvement of Competitiveness (EKVAN).

The objective of this programme is the provision of necessary financing for the development of high-tech systems or high value added services. The beneficiaries of these funds are companies, universities or public research centres.

To date, a total of 84 large scale projects have been approved within this research consortium amounting to a total of 26 billion drachmas. The execution period is from 1995 to 1997. Supervision is carried out simultaneously in order to ensure that the objectives proposed at the beginning of the cooperation agreement are achieved.

One of the objectives of the Greek Government when establishing the best policy to be carried out in relation to R&D is that of the decentralisation of activities in this area and endowing regions located far from the large urban centres with infrastructure and financing sufficient to be able to conduct scientific research projects of whatever type. Among the concrete actions envisaged in Greek regions within this programme, that of EKVAN is worth noting.

23 projects have been approved under this programme with a total budget of 3.2 billion drachmae. A fundamental condition of participation in the programme was that 70% of the activity be carried out in the regions of Macedonia, Thrace, Epirus and the North Aegean. As already indicated, the objective is the promotion of research and technology in peripheral territories located far from the evident concentration in connection with the capital Athens. The approved projects cover a large number of areas but concentrate mainly on environment, agriculture, health, computer science, materials, and technology culture.

Based on the distribution of the projects, 37% of financing was absorbed by individual companies, while 43% is spent by universities and public research centres. The remainder came from other institutions, both public and private.

Analysis of the participation by region, shows that Epirus (25%) and Thrace (20%) are those which have been most involved in this programme. As regards sectors, the highest number of proposals hailed from the areas of agriculture (29%), environment (22%) and data processing (21%).

Regional Technology Programme for Central Macedonia

The Minister for Macedonia and Thrace set up the Committee of the Technology Programme for Central Macedonia in 1995. In comparison with other types of programmes, this programme does not finance any type of activity related to R&D, such as subsidies for concrete research projects, financing for the construction of infrastructure nor coordination of projects carried out by public research centres or universities. Rather it aims at increasing company demand for technology in the region by means of studies and audits determining technology needs and analysing how these could improve production or services. Once these needs have been determined, it is possible to elaborate which technologies or innovative processes are needed in the future, and include relevant actions in existing programmes which provide solutions to these problems. This eliminates the need to create new programmes.

The budget estimated for this programme is 400,000 ECUs. The principal activities to be carried out range from making companies or public organisations aware of themes relating to technology to more technical areas such as the realisation of technical audits in companies from different sectors, competition analysis, the incorporation of foreign specialists, etc.

Networks of Human Mobility for the Dissemination of R&D Know-How

At the beginning of 1995, the GSRT initiated a specific action for the dissemination of scientific knowledge among personnel employed in R&D related tasks.

The principal objective of this action is the improvement of communication links at the same time creating new channels of communication among researchers in general and executive personnel within companies or organisations in some way involved in research. This enabled the elimination of one of the structural problems of the Greek science and technology system, namely the lack of links or isolation existing among the different units or organisations responsible for research and development.

The achievement of this objective will have a twofold effect:

1. To bring the scientific environment closer to the real needs of enterprise thus eliminating the present situation of science being stimulated only by public programmes and projects.

2. To encourage a multilateral or multidisciplinary approach to the principal socio-economic problems suffered by the country.

These networks are made up of researchers and professionals from third-level education, public research centres and enterprise, which in some way share common interests in the field of science. They also try to find a solution to socio-economic problems, perhaps through different approaches, or are interested in the exchange of scientific or technological services.

The organisation of the network is initially unrestricted although it is proposed that a representative number exist both in specific sector of industry and in each region of Greece. These networks will enable close relationships with other European networks of information exchange, securing the internationalisation of Greek scientific research.

The minimum number of participants in each network is 10 researchers and the maximum, 30. The participation of an end-user of know-how, in this case an enterprise or public research centre, is also necessary.

The duration of the project is two years (1995-1996) and the budget assigned to 1995 activities reached a total of 300 million drachmae. GSRT participation ranges from 50 to 70% of the total budgeted in each individual network.

Programme for the Determination and Assimilation of Standard Practices of New Technologies

In recent years, Greece has made great efforts to develop new technologies and exploit research results from the production sector. It becomes clear, through analysis of how this has been carried out, that the majority of the new technologies or innovative processes adopted originate from the results of research processes conducted in other companies, especially those of high technological content. This is not to say that there has been a breach of copyright nor have the general rules of competition been violated, as these tend toward free distribution processes, but simply that technology has been acquired paying the corresponding fees. This is not the problem, but rather the inappropriate use made of the technology acquired. The adaptation of technology to the particular needs of the company which has acquired it is not always an easy task and is sometimes even impossible. On the other hand, it is clear that no company, no matter how large, can consider itself self-sufficient regarding the development of new organisational processes or technologies.

Faced with these two facts, the GSRT began a specific programme for the dissemination of the best international practices regarding the use of technological advances. The object of this is to make the Greek productive sector aware of standard practices concerning technology in companies of the same sector, both national and international, and how they can adapt such new technologies to their specific needs.

It could be argued that this type of activity gives rise to an increase in technology imports which in general terms is not positive. However this statement is only applicable in those countries with a high level of technological development. This is not the case of Greece. Here the acquisition of foreign technology is more favourable in that it serves as a driving force or spur for the future development of new domestic technological processes. According to data provided by the GSRT, it is foreseen that technological transfer achieved from abroad thanks to this programme will double, and in future years, will lead to a more balanced situation.

The programme budget is 1 billion drachmae in total. Contribution by enterprise is expected to reach 2-3 billion drachmae.

Enterprise may expect the following benefits:

- Greater awareness of own achievements concerning research.
- Development of new evaluation methods of R&D activities
- Acquisition of new technologies, especially those relating to the company activity.
- Knowledge of the competitors' performance as regards technological development.
- Knowledge of the benchmarks being used by competitors.
- Knowledge of the operative methods of other leading foreign companies.

The company must decide the level that it wishes to reach regarding technological transfer, although it is encouraged to use its own previous positive experiences in order to determine its position.

Overall the programme offers:

- Aid to the company when determining the market segment in which to compete and develop.
- Collaboration in establishing on-going internal training programmes.
- Collaboration in the strategic planning of the company in order to reach a leading position in the market.
- New techniques for organisational restructuring.
- Aid for effective management of the new technologies incorporated.

Evaluation of the Results of Public Programmes

As will be seen in the chapter dedicated to the analysis of the principal indicators of the S&T system in Greece, public R&D spending is the largest entry within total national spending on scientific research. It is therefore fundamental to conduct an, albeit brief, analysis of public research programmes developed. It is still early days to evaluate the development of the EPET II Programme, integrated into the Community R&D Support Framework, which will end in 1999. Therefore, we will concentrate on the previous programme, EPET I, and on the EU initiative, STRIDE. Although both have finished, they will serve as an example of the development of public programmes in Greece.

The STRIDE programme (Science and Technology for Regional Innovation and Development in Europe) is a programme of the European Union. Its aim is to provide assistance to national operational programmes related to science and technology. The European Commission launched this initiative in 1990 with the clear objective of obtaining the convergence of the different scientific and technological systems of the less favoured Member States, these being the entire territories of Greece, Portugal and Ireland in addition to certain regions of the other members. Financing was in part from the Structural Funds of the European Union and in part from the governments involved in the activities.

The first phase of STRIDE had an anticipated length of three years (1990 - 1993) and financing more than 70 million ECUs. A second three year phase was initiated in 1993. In general financing is provided by the European Commission through the Structural Funds whose principal mission is that of encouraging inter-territorial cohesion within the Union. Remaining financing originates from the Greek government. At present the activities of STRIDE are incorporated into EPET II.

Based on the conclusions of the Evaluation Committee established to appraise the results of EPET I and STRIDE, these must be classified as satisfactory in terms of both execution and financing. The most important conclusion was that the development of these programmes lead to the establishment of optimum conditions for the future development of the Greek S&T system, thus enabling the system to reach the same level as neighbouring countries. This new condition supposed that the isolation of the different units responsible for development of scientific research had been eliminated and that adequate communication links had been established between the scientific community and production. If so, Greece would be in a position of carrying out competitive research, albeit lagging behind its competitors by some years.

In particular 57% of the projects under the EPET I and STRIDE initiatives achieved their objectives, while only 10% experienced noteworthy deviations with respect to the original idea. The remaining 33% achieved their basic objectives, but with a certain degree of deviation (see Table 3-4).

Table 3-4: EPET I & STRIDE Programmes: Results

Objective Totally Reached	57%
Objective Partially Reached	33%
Objective Not Reached	10%

Source: GSRT, 1996.

The reason for such a high number of projects not achieving the anticipated results is due to a number of causes. Perhaps the most important is the vague definition of principles or results when deciding on participation in public R&D programmes. Another reason may be the resulting scarce economic potential. Projects with principles and objectives were clearly defined from the beginning, in relation to technical and administrative aspects as well as to the contribution that they could make to the national socio-economic background, and were accomplished satisfactorily.

As regards financial auditing of the projects, a wide majority of them showed acceptable financial administration. Only a small percentage showed levels of subcontracting considered to be excessive, presented signs of mismanagement deemed to be unacceptable by the promoting public entities. As is logical in these cases, the devolution of funds was demanded; to date this only reaches 1% of funds assigned to the different projects. To be specific, of the total budget, 15 billion drachmae, only 150 million drachmae has been reclaimed. This figure should be considered low, indicating an efficient control system, at least administrative, of the activities carried out by the Greek public S&T system.

As already noted, the quality of results have served as a springboard to achieve the desired level of competitiveness in line with other countries with similar backgrounds. In general, they can be considered satisfactory. The prestige achieved at European level by various research units would not have been possible without considerable support on the part of public administration nor the pre-definition of priority research lines taking into account the principles of national interest.

The General Secretariat for Research and Technology has created a Project Evaluation Register in which the experiences resulting from the development of EPET I and STRIDE will be recorded. This register will serve as a source of consultancy in the course of EPET II to accept or reject new proposals formulated in the calls for application of the programme initiatives.

This brief evaluation of the activities accomplished under the Community R&D Support Framework is based on the conclusions obtained by 47 appraisal committees, consisting of international experts, GSRT staff, foreign audit companies and the Advisory Boards of the different operational programmes. These conclusions were presented in June of 1995 in the Pasteur Institute.

The following table summarises the results of the first Community R&D Support Framework in Greece.

Table 3-5: Evaluation of the Results of EPET I & STRIDE

	Proportion of Projects (%)		
	EPET I	STRIDE	Total
Totally successful	38	18	29
Successful	42	37	37
Successful with deficiencies regarding its aims	21	38	29
Unsuccessful	0	13	6

Source: GSRT, 1996.

PUBLIC EXPENDITURE ON SCIENTIFIC AND TECHNOLOGICAL RESEARCH

In comparison with other countries whose S&T systems are more developed, the role of the public sector in Greece is fundamental when establishing and promoting the main lines of scientific research. In the following section, the breakdown of Greek public spending will be outlined, with the exception of the general indicators of spending already provided.

Public spending on R&D has increased spectacularly in recent years. It is sufficient to point out that in 1993 this was 72,965 million drachmae compared to 43,974 million in 1991. This supposes an annual growth of 65.9% measured in terms of current drachmas. If we discount inflation (one must take into account that the Greek level of inflation is considerable if compared with other EU countries), this increase supposes an annual percentage of 27.9% (constant prices). If we consider the evolution of previous years, the rate of increase reached 28.8% and 13.1% respectively.

When making reference to R&D public spending, that referred to is the sum of the different budgets earmarked for state research funding, independent of the origin of these funds. In this section the budgets of public research centres, university and professional training institutes (Technological Training Institutes) are included. The evolution of the gross quantity employed does not provide a clear idea of how the Greek public science system is evolving, given that such spending overly depends on the size of the Greek economy. A suitable indicator reflecting the endeavour of public administration is the quotient between public expenditure on scientific and technological research (PESTR) and GDP. These relations can be observed in Table 3-6.

Table 3-6: Public Expenditure on Scientific and Technological Research (PESTR) (Millions of Drachmae)

Year	Current prices	Constant prices	PESTR / GDP
1986	13,085	13,085	0.23
1988	19,814	15,565	0.17
1989	31,852	21,846	0.29
1991	43,974	21,030	0.28
1993	72,965	26,905	0.35

Source: GSRT, 1996.

The most significant data that can be extracted from the above table is that the fraction of GDP intended for R&D saw a sharp increase at the end of the 80's, followed by three years when it remained steady. At the end of this period, it again increased to finally reach 0.35%.

Table 3-7, in which the evolution of the origin of R&D funds are outlined, shows the reasons behind this substantial increase in the investment level. It is possible to see which sections have increased and which have been maintained.

Table 3-7 Analysis of Public Investment in R&D According to Funding Source (Millions of Drachmae).

Funding source	1991 Current Prices	1993 Current Prices	1993 Constant Prices 1991
Government Budget & Other Public Funding Sources	32,652	44,610	34,395
Self financing of institutions by exploitation of resources	1,357	3,545	2,733
Public sector companies	937	1,236	953
Private companies	530	720	555
EU Framework Programme	2,791	11,376	8,771
EU Structural Programmes	5,562	11,044	8,515
Other Sources	155	434	335
Total	43,974	72,965	56,257

Source: GSRT, 1996.

It should be pointed out that "Other Sources" refers to quantities whose origin is not directly linked to central Administration. For the most part these quantities come from the regional administrations. Though in previous chapters we have indicated that large part of the funds intended for scientific research originate from the European Union, this is not going against those indicated in the table in which government spending appears as a fundamental entry. Although the programmes included in this budget figure as government spending they originate from EU funds.

Various conclusions can be drawn from this table. First, the financial injection from the European Union through the R&D Framework Programme quadrupled in the period 1991-1993. This means an annual increase of 101% in current drachmas and 77.3% in constant drachmas. This data is significant given that the R&D Framework Programme of the European Union is not a cohesion instrument and therefore demonstrates the growing prestige of Greek institutions in the area of scientific research. Undoubtedly if this had not been so, funds destined by the European Commission may have been frozen or reduced. The data also indicates that the Greek S&T system is moving forward in the same way and with the same interest as the other EU members. This indicates gradual integration and convergence with the structures of this organisation.

With respect to structural programmes, which do have as objective, regional cohesion within the EU, the Greek State achieved an increase in the level of funds assigned from 5,562 million drachmas in 1991 to 11,044 million in 1993. This translates into an increase of 98.6% in current prices and 53.1% in constant prices. Comparison of these figures with the evolution of a temporarily wider series, would correspond to figures in the area of 41% and 24% respectively. The difference can be explained by the fact that the main part of funds from the STRIDE and EPET I programmes were received in 1991.

Perhaps the most negative data of this table is the level of private sector funding. The total outlay related to R&D reached a figure of 1,956 million drachmae in 1993 and 1,457 in 1991, an increase of 34.2% in terms of current prices and 3.2% in constant prices. Funding by the private sector, as reflected in the table, through participation in public R&D programmes has been low. The increase is only 15.9% or 1.7% respectively. The same can be seen in the comparison of the contribution of public companies which increased from 937 to 953 million drachmae in constant prices and which is again a low figure. Several of the reasons justifying this situation are outlined below but in general they show a marked distance between public scientific research plans and the Greek productive sector.

Increases in central government financing were also modest. As can be appreciated in Table 3-8, spending increased from 32,652 million drachmae in 1991 to 44,610 in 1993. Initially this means an increase of 36.6% in only 3 years but comparison in constant prices, a more accurate indicator of government endeavour in strengthening R&D activities, shows that the increase is 5.3%. The total figure of 44,610 million drachmae can be broken down as follows:

Table 3-8: Central Government Budget on R&D in 1993.
            (Millions of Drachmae)

	1993
Public budget	43,354
Other public resources	1,256

Source: GSRT, 1996

Institutions with own financing in addition to other external funds, apart from the funds provided by the EU, have experienced spectacular increases. The rate of variation is 161% in current prices. However as these figures do not surpass 4 billion drachmae, they do not greatly affect the global figure of national investment in scientific research.

Analysis of the table gives rise to the general conclusion that the public authorities, in spite of not making efforts in promoting R&D activities, do, however, know how to take advantage of the resources being made available by the EU.

The percentage of state participation in the financing of R&D can be observed in Table 3-9.

Table 3-9: Analysis of Public Financing for R&D
            (Percentage)

Funding source	1991	1993
Government Budget & Other Public Funding Sources	74.24	61.14
Self financing of institutions by exploitation of resources	3.09	4.86
Public sector companies	2.13	1.69
Private companies	1.18	0.99
EU Framework Programme	6.35	15.59
EU Structural Programmes	12.65	15.14
Other Sources	0.35	0.59
Total External Capital	19.35	31.32
Total	100%	100%

Source: GSRT, 1996.

It is clear that EU funds are increasing in importance with respect to the overall total invested. Foreign sources increased in the period 1991-1993 from 19.35% to 31.32% meaning that state participation has shrunk from 80.7% to 68.7%.

With respect to public institutions where research is carried out (public research centres and universities mainly) the breakdown of spending can be seen in Table 3-10.

Table 3-10: Analysis of Public Budget on R&D by Sector 1993

Funding source	Public Research Centres		Third Level Education	
	Millions of Drachmae	%	Millions of Drachmae	%
Government Budget & Other Public Funding Sources	199,907	77.7	5.703,	22.3
AEI Budgets	0	0.0	17,744	100.0
Other public sources	533	42.4	723	57.6
Self financing of institutions by exploitation of resources	1,122	31.6	2,423	68.4
Public sector companies	244	19.7	992	80.3
Private companies	161	22.4	559	70.6
EU Framework Programme	4,165	36.6	7,211	63.4
EU Structural Programmes	5,698	51.6	5,346	4,834
Other Sources	294	67.7	140	32.3
Total	32,124	44.0	40,841	56.0

Source: GSRT, 1996

Considering public institutions as a whole, third level education is at the head of the table in terms of R&D spending and accounts for 56% of total invested, some 40,841 million drachmae. However there also exists a noteworthy discrepancy in the number of researchers in each group. More than 6,500 researchers carry out their activities in universities or institutes of higher education while in the rest of the public organisations the figure does not reach 2,600.

The investment figures are not directly comparable, given that they are based on different methods of calculation. Data provided for the calculation of funding for universities and third level education research institutions is not exactly the total of transferred funds, but rather an estimate of expenditure according to the method of the so called "weight coefficients" in third level education. The figure given as "public budget" refers exclusively to team participation within these organisations in R&D programmes financed by the State. In spite of the figures not being directly comparable, the table is included with the objective of showing that the majority of resources are to finance activities in higher education. In fact, such activities attracted 80.3% and 70.6% of government expenditure and private sector financing compared with 19.7% and 22.4% destined for public research centres. This large difference (4/1 and 3/1 respectively) demonstrates clearly the differences that exist.

Examination of financing by the EU Framework Programme shows that the situation is clearly in favour of third level education, assuming 63.4% of total expenses of projects. Public research centres, on the other hand, benefit greatly from the ESF, the percentage of funding approaching 52%. The reason for such a marked difference with respect to the Structural Funds is due to the fact that EPET I comes within the framework of these funds as does EPET II. The first phase of the former placed emphasis on the renovation of the infrastructure of public research centres which means heavy investment.

The only real disadvantage of the third level education sector in comparison to public research centres is due to the fact that funding from external sources has increased considerably in recent years. In fact universities and related centres only receive 32.3% of the total, due mainly to contracts with international organisations, which in 1993 was greater than 170 million drachmae.

The regional distribution of public spending on science and technology research can be broken down as shown in Table 3-11.

Table 3-11: Regional Distribution of Public Expenditure on Science & Technology Research

Region	1993 (Millions of Drachmae)	1993 %	1991 (Millions of Drachmae)	1991 %
Eastern Macedonia, Thrace	3,035	4.16	1,785	4.06
Central Macedonia	14,495	19.86	8,198	18.64
Western Macedona	131	0.18	171	0.39
Thessaly	1,125	1.54	531	1.21
Epirus	2,042	2.80	1,063	2.42
Ionian Islands	285	0.39	142	0.32
Western Greece	5,703	7.82	3,097	7.04
Central Greece	292	0.40	111	0.25
Peloponnese	188	0.26	338	0.77
Attica	35,469	48.61	22,185	50.45
Northern Aegean	724	0.99	203	0.46
Southern Aegean	1,089	1.49	185	0.43
Crete	8,388	11.50	5,964	13.56
Total	72,965	100	43,975	100

Source: GSRT, 1996

The first conclusion to be drawn from this table is the intense centralisation of resources in the region of Attica. The principal reason for this is the vast number of public research centres, universities and important companies located in four regions, one of which is Attica.

In fact, in 1993, approximately 48.6% of public spending was absorbed by the region of Attica, 19.68% by Central Macedonia, 11.5% by Crete and 7.82% by Western Greece. The remaining regions obtained only 12.1%. Certain decentralisation with respect to the evolution of these indicators can be observed in 1991-1993, but the difference is so insignificant that the cause may be put down to circumstantial reasons rather than to an effective policy of resource redistribution.

Regions such as Central Greece and Peloponnese are to be found at the other extreme. In such regions, the quantities are extremely low with respect to the rest

Table 3-12: Regional Distribution of Public Expenditure on Science & Technology

Region	Public research Centres		Third level education	
	1993 (Millions of Drachmae)	1993 %	1991 (Millions of Drachmae)	1991 %
Eastern Macedonia, Thrace	1,121	3.49	1,914	4.69
Central Macedonia	3,495	10.88	10,999	26.93
West Macedonia	124	0.39	7	0.02
Thessaly	517	1.61	608	1.49
Epirus	276	0.86	1,766	4.32
Ionian Islands	155	0.48	130	0.32
Western Greece	1,990	6.19	3,713	9.09
Central Greece	292	0.91	0	0.00
Peloponnese	188	0.59	0	0.00
Attica	16,687	51.95	18,782	45.99
Northern Aegean	196	0.61	258	1.29
Southern Aegean	737	2.29	352	0.86
Crete	6,346	19.75	2,042	5.00
Total	32,124	100.00	40,841	100.00

Source: GSRT, 1996

Table 3-13: Regional Distribution of Public Expenditure on Science & Technology: Comparison of Subsectors & Years

Region	Public research Centres		Third level education	
	1993 (Millions of Drachmae)	1993 %	1991 (Millions of Drachmae)	1991 %
Eastern Macedonia, Thrace	4.13	3.49	3.97	4.69
Central Macedonia	10.02	10.88	20.89	26.93
Western Macedonia l	0.71	0.39	0.00	0.02
Thessaly	1.13	1.61	1.30	1.49
Epirus	0.48	0.86	4.72	4.32
Ionian Islands	0.07	0.48	0.62	0.32
Western Greece	5.59	6.19	8.77	9.09
Central Greece	0.47	0.91	0.00	0.00
Peloponnese	1.41	0.59	0.00	0.00
Attica	56.00	51.95	43.86	45.99
Northern Aegean	0.02	0.61	0.99	1.29
Southern Aegean	0.36	2.29	0.51	0.86
Crete	19.61	19.75	6.37	5.00
Total	100.00	100.00	100.00	100.00

Source: GSRT, 1996

of the country although this is of little surprise given that neither of the two regions mentioned have any significant concentration of public research.

The regional distribution of public research centres, universities and related institutions is shown in Table 3-12 and in Table 3-13. The conclusions that can be drawn from this data are various. The first is that the concentration of resources in the region of Attica is especially significant with respect to third level education and not so much in the case of public research centres. Furthermore if data regarding the latter is compared with that of 1991, a fall of approximately 4% can be observed. Universities show contradictory trends, that is to say, increasing spending. As indicated previously, centralisation of spending is especially noteworthy in four specific regions (Attica, Central Macedonia, Western Greece and Crete), which absorb approximately 89% of the resources intended for public research centres and a similar fraction of resources intended for universities and related centres.

Special attention should be paid to the case of Crete which is to be found in second place as regards the level of investment, having almost double the resources of Central Macedonia and almost the triple that of Western Greece. This has been achieved due to the FORTH network (Foundation for Science & Technology).

FOURTH R&D FRAMEWORK PROGRAMME

This section describes Greek participation to date in the IV R&D Framework Programme of the European Union. The aim of this description is twofold: on the one hand to outline Greek participation, that is to say, identity the programmes with Greek representation and compare this participation with that of other Member States; on the other, provide an accurate reflection of how the general guidelines of the Greek science and technology system are performing. As has been repeated on numerous occasions throughout this book, the purpose of the R&D Framework Programme is not that of interterritorial cohesion (other mechanisms such as the ESF, etc. exist for this reason) but rather the development of new technologies, products or innovative processes that improve the socio-economic context of the EU as a whole. Therefore financing is not guided by the criteria of balanced distribution of resources. Projects will be awarded to those working groups or institutions best suited to carrying them out or benefiting from them.

In the phases previous to project evaluation, Greece's position was outstanding in comparison with previous calls, participating in 6% of the total number of projects. This figure may be considered noteworthy taking into account that Greek research activities represent 0.6% of the EU total.

With respect to the Advanced Communication Technologies and Services programme (ACTS), Greece participates in 5.6% of the total of selected proposals. Participation by other countries can be observed in Table 3-14.

Distribution in the Telematics Applications programme is similar, although here Greek participation is even greater, reaching 7.6% of the total. A breakdown of this figure shows that significant participation corresponds to the areas of environment (17.3%), health (8.3%), transport (8%) and administration (7.7%).

Greek participation in the Language Technology programme is 6% with good possibilities of this increasing further.

It should be noted that extensive effort has been made by the Greek state in order to encourage the participation of research teams in specific areas of these programmes, both to develop new technologies of national interest as well as to achieve wide technology transfer.

Figure 3-4: Country Participation in the ACTS Programme

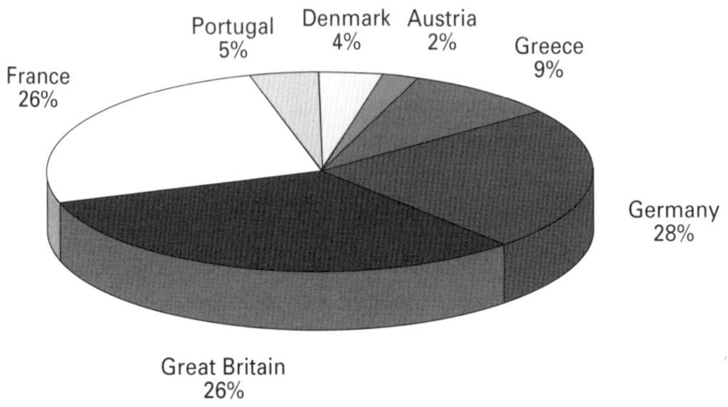

- Portugal 5%
- Denmark 4%
- Austria 2%
- Greece 9%
- France 26%
- Germany 28%
- Great Britain 26%

Source: GSRT, 1996

The objective of this brief summary of data is not so much to offer a detailed description of Greek participation in the IV Framework Programme, but rather to show that the level of competition of public research centres in Greece is high in comparison with the country's overall technological contribution to the EU. This data is highly positive given that it places Greece in an excellent position to reach the competitive level of the principle, greater developed countries of the EU. This will become reality when R&D financing rates are increased and the main structural problems of the science and technology system are solved.

R&D IN THE PRIVATE SECTOR

In recent years the Greek economy has evolved in a manner different to that of its EU neighbours. The Greek economy has a high public sector presence which in 1994 cornered more than 24% of investment. Evidently the country should be considered developed, but the reality is that along with Portugal, these countries are the two least developed of the European Union. However, while Portugal is presently undergoing a rapid process of convergence with the EU average, Greece is diverging.

In the case of production, as in the rest of the Member States, traditional sectors such as agriculture, fisheries or mining are being abandoned in order to change over to economies based on the services and industry sector, which are becoming increasingly more important.

Table 3-15: GDP by Sector
(1993)

Sector	Billion drachmae
Agriculture and Fisheries	1,983
Mining	173
Manufacturing	2,220
Electricity, Gas, Water	377
Construction	946
Transport and Communication	1,016
Trade	1,975
Banks and Financial Services	446
Other services	5,287
GDP at factor cost	14,423

Source: The Economist intelligence Unit, 1996.

Greece is one of the poorest countries of the European Union. GDP per capita is approximately 35% that of the Union average. Nevertheless, the standard of living in the principal urban nuclei comes quite close to that of the rest of the countries of the same background. The services sector as well as that of industry have been set up around such areas, especially in the metropolitan zone of Athens and in the regions of Attica, Central Macedonia, Crete and Western Greece. In the rest of the country, the presence of industry is much less and GDP per capita is only 20% of the average of the European Union.

In short, with reference to the state of Greek industry, perhaps the most important fact is the existence of a production sector less developed than that of the other countries of the environs and that the public sector has acquired a fundamental importance in the socio-economic framework of the country.

Greek industry has shown in the last years, as have the other less developed countries of the EU, low capital intensity and high intensity in human resources. This gives rise to the understanding that investment in the development of new technological processes is scarce. It is therefore logical to suppose that private Greek science and technology depend basically on the acquisition of technology from abroad due to the scarcity of domestic research.

Before entering into a description of the basic parameters that mark the technological development of Greek industry, it should be mentioned that a large number of the companies are very small (more than 93% have less than 8 employees), and almost all of them are no bigger than the size of an SME (more than 99.5% have less than 50 employees). The size of companies is fundamental when considering scientific research. Evidently, it is impossible that companies with scarce resources be involved in R&D processes and therefore, unless the government can facilitate actions through consortia consisting of various companies, technological development will be obliged to depend on the acquisition of technology.

Examination of recent reports of the General Secretariat for Research and Technology, makes possible the assertion that the contribution of the private sector to financing of the Greek S&T system is "hopelessly low". Such a situation is due to a series of circumstances, some of which have already been mentioned:

- Small size of companies.
- High interest rates.
- High dependency on the acquisition of technology from abroad.
- Public investment in the private R&D sector combined with lack of evaluation and follow-up of results.

Keeping in mind these same reports there are other reasons, perhaps of greater relevance, which could also be pointed out, such as lack of confidence on the part of the business sector in the competence of Greek researchers to tackle the areas of research necessary in order to develop new products or appropriate innovative processes. Against this background, and with levels of private investment that have not shown noteworthy growth in recent years, public sector dependency with regard to R&D is becoming increasingly important. So much so that if it were not for this, convergence with the rest of the EU countries would be practically impossible.

Moving on to the most important indicators regarding R&D activities within the Greek private sector, the main entries of company spending on R&D can be seen in Tables 3-16 and 3-17 according to the type of research, activity and funding source.

Table 3-16: Industry Expenditure on R&D in 1991
(Constant Prices - 1990 : Millions of Dollars)

Type of Economic Activity	$	%
Agriculture and Fishing	0.3	-
Mining	5.5	5.8
Manufacturing	57.6	61.4
Services	28.1	29.9
Others	2.3	2.4

Source: OECD, 1995.

As can be seen, the sectors absorbing the greater quantity of resources are those of services and industry. This fact clearly demonstrates that both sectors are the most important in the Greek economy while others, such as mining, agriculture or fishing, do not reach 6% of total investment. This displacement of investment is occurring all over Europe, as indicated throughout this book, but in Greece the change is more marked. In other countries, such as Spain or France, not only has the competitiveness of primary sectors areas not been pushed to one side, but an attempt at revitalisation has been made in order to increase their productivity.

Table 3-17: Sources of R&D Financing in the Greek Industry Sector 1993

Source of Financing	Millions of Drachmae	%
Administration	1,878	6.3
Own Funds	19,063	64.1
Private Non-Profit Making Organisations	-	-
Universities	-	-
Foreign	8,798	29.5
Total	29,739	100.0

Source: OECD, 1995.

Another important aspect to be underlined is that of scarce public financing received by companies to develop R&D activities. Of the total invested, only 6.3% comes from administration (in countries like Portugal the percentage is similar), while in other countries (Spain, France, Italy, etc.) it is around 20%.

Table 3-18 shows spending on R&D according to the activity carried out. As can be seen, and as has been mentioned at the beginning of this chapter, the services sector is daily gaining greater importance and only the industrial sector as a whole surpasses it. Other sectors such as electricity, construction, etc. provide less support.

The most important conclusion that can be reached on analysis of Table 3-18 is the degree of fluctuation in R&D investment in the business sector. In comparison with other EU Member States, experiencing a process of convergence with their EU partners and therefore showing increasing R&D values (even notably higher than the normal growth of the sector), there is no evidence in Greece of the willingness of enterprise to develop R&D processes. This is due to various reasons, two of which are indicated below:

- Overconfidence in the role of public administration in financing this type of activity.
- Lack of awareness of the Greek businessman regarding R&D.

With respect to the number of researchers, the total of full-time researchers in the business sector in 1991 was not greater than 2,330. This figure cannot be considered high, especially if compared with the total number of researchers nation-wide. For example, in the university sector the total number was 4,170

Table 3-18: R&D Expenditure in the Business Sector According to Activity
(Millions of Dollars - 1990 Prices)

Sector	1988	1989	1991
Agriculture	5.6	4.8	0.3
Mining	2.4	1.2	5.5
Industrial	56.4	65.1	57.6
Food	2.1	2.2	5.1
Textiles	1.1	0.6	0.5
Wood, paper	0.1	0.4	1.0
Oil, Chemical	- -	- -	- -
Non-metal mineral products	2.8	3.3	3.9
Metals	2.5	3.3	0.8
Metal products	12.3	14.4	5.3
Machinery, Equipment	25.4	30.7	25.1
Other Manufacturing	0.1	0.0	1.4
Recycling	- -	- -	- -
Electricity, Gas, Hydraulic Resources	2.9	1.8	2.0
Construction	0	0	0.3
Services Sector	7.8	5.6	28.1
Sales, trade	- -	- -	0.1
Hotels, Restaurants	- -	- -	- -
Communications	0.0	0.0	0.5
Transport and storage	0.2	0.3	- -
Financial Intermediation	- -	- -	2.3
General financial activities	- -	- -	25.0
Communication, society	- -	- -	0.1
Total	75.1	78.5	93.8

Source: OECD, 1995.

Table 3-19: R&D Personnel in Industry
(Full-Time)

Sector	1988	1991
Agriculture	171	11
Mining	55	78
Industry	1,193	1,288
Services	284	830
Other	63	37
Total	1,766	2,244

Source: OECD, 1995.

in 1991 and 6,768 in 1993, a figure nearly double that of 1991. This is due to excessive dependency in R&D on the public sector.

The solution to these problems, which could basically be classified as structural, can be found in a change of R&D policy. Obviously these changes will not be produced if the circumstances are not favourable. It is necessary to increase financing of these activities, but this is not the only determining factor. Links among the entrepreneurial sector, public research centres and universities must be significantly tightened. At present it is not conceivable that the public sector deploy their means in order to develop new innovative processes without taking into account that one of their missions is that of activity coordinator. In addition, the needs of the Greek businessman cannot be satisfied solely through pure R&D activities. It is essential to develop technical support that can provide industry with the appropriate know-how. In Greece, this support has not yet been developed and is one of the points that must be strengthened. Greek public administration has made available programmes to industry with the specific objective of supporting research areas related to production (EPET II, Subprogramme 2). The companies must know how to avail of such programmes.

MAIN R&D INDICATORS

As will be seen throughout this section, Greece is endeavouring to converge with the other countries of its relatively irregular environment, depending on those sectors to which reference is made, in the areas of science and technology. Since joining the European Union and in the face of increasingly competitive markets, the financial injection received from the European Commission has been considerable for the modernisation of the productive and economic structure as a whole. Although much has been achieved and the differences separating the country from the other Member States have been reduced, Greece is still the least developed country in terms of R&D.

Table 3-20: **Expenditure on R&D in the period 1988 - 1993**

Year	Millions of drachmas (current)
1988	27,589.7
1989	40,998.1
1991	59,503.0
1993	100,862.0

Source: OECD, 1995.

As can be seen, the level of investment in a five year period was practically quadrupled at the end of the 80's and beginning of the 90's. Initially this could be seen as optimistic, but reference to these same quantities in constant prices, shows that expenditure on R&D has only doubled. Comparison of these figures with Spain or Portugal, countries also involved in processes of convergence with the community average, show that growth in the same period was close to a factor of four. In addition, the differences separating Greece from the other countries in the OECD are large. In international terms, investment realised is not comparable, not only with larger countries, but also with those of a smaller size.

If we compare the expenditure on R&D as a percentage of GDP the differences are even greater. The evolution of this figure in Greece and its comparison with other countries is shown in Figure 3-5 and in Tables 3-21 and 3-22.

Figure 3-5: Expenditure on R&D as a percentage of GDP (%)

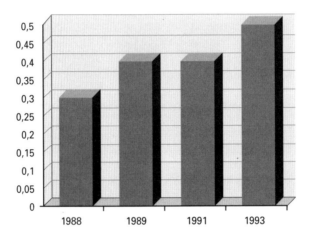

Source: GSRT, 1996.

Table 3-21: Comparison of R&D Expenditure as a Percentage of GDP in Comparison with EU Countries

Country	1989	1990	1991	1992
Germany	2.88	2.81	2.83	2.81
France	2.34	2.40	2.42	2.45
EU(mean)	2	2.01	2.1	2.3
Greece	0.3	0.4	0.4	0.46
Spain	0.75	0.81	0.87	0.92
Italy	1.24	1.35	1.35	1.4
Portugal	0.5	0.6	0.65	0.7

Source: National Statistic Institute (Spain), OECD.

As can be observed, Greece is located well below even the other less developed countries of the EU. Spain, as well as Portugal, reached levels higher than 0.5% of GDP while Greece did not reach this figure. If compared to the EU average, the difference is even more significant, reaching almost two percent.

The best way of appreciating at a single glance the effort being made by a country in the area of R&D is to illustrate the trend in the growth of R&D spending in comparison with the net growth of the economy, measured in terms of the Gross Domestic Product. The case of Greece can be seen in Figure 3-6.

Figure 3-6: Breakdown of the recent evolution of expenditure on R&D

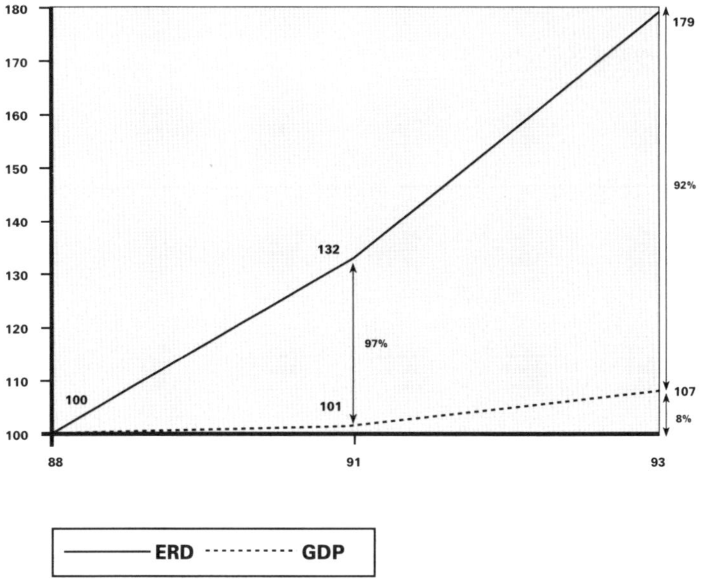

Source: OECD, 1995
Own Elaboration

Taking 1988 as base with a value of 100 common to both GDP as well as R&D spending, the latter acquires a value of 132 in 1991 while GDP had only a value of 101. In 1993 the difference is still greater, expenditure on R&D being 179, while GDP remained at 107. Further analysis of the graph shows that the growth in investment in scientific matters can be divided in two; that inherent to the growth of the economy, and that derived from a greater encouragement of investment in scientific research activities. In 1991 the growth of R&D expenditure was due in 3% to the growth of GDP and in 97% to the increase in investment. This situation is no better in 1993, where 8% may be attributed to the GDP and 92% to increased investment. In other words, the institutional impulse is going above and beyond pure economic growth.

This increase in effort is optimistic. It is sufficient to take into account that in spite of such a slight increase in GDP, investment in R&D has been significant, thus denoting an increase in interest with respect to scientific research processes.

As previously indicated, the weight of R&D in Greece is sustained for the most part by public administration, especially through the financial injection of the European Union. This had become one of the principal structural problems of research in Greece. It can only be explained by the fact that public sector presence is very important in the economic model of the country and that there is a low level of awareness among Greek entrepreneurs regarding R&D related matters. As can be seen from the following table, there are substantial differences in the origin of funds (public, private, foreign) in Greece with respect to the rest of the industrialised countries

Table 3-22: Funding of scientific research

Country	GDP (M $)	R&D Expenditure (% of GDP)	Industry Funding	Government Funding
USA	149,225.0	2.77	50.6	47.1
Japan	62,865.0	2.88	77.9	16.1
Germany	31,585.3	2.73	62.0	35.1
France	23,768.4	2.42	43.5	48.3
England	20,178.3	2.22	49.5	35.8
Italy	11,964.3	1.30	47.3	51.5
Greece	336.3	0.45	19.4	68.9
Portugal	501.8	0.61	27.0	61.8
Spain	3,888.8	0.85	47.4	45.1
Belgium	2,751.5	1.69	70.4	27.6
Austria	17,96.7	1.40	53.2	44.3
Finland	15,41.8	1.87	62.2	35.5

Source: EAS, 1991

The imbalance existing in the origin of funds for scientific research may be appreciated in Table 3-23:

Except in the case of Portugal, the rest of the countries present investment values of 50% or higher with reference to the percentage of financing by the productive sector. In Greece this value only reaches 20%. The same can be deduced from Graph 1-7. The graph shows percentage spending on R&D by sector. If the percentage used by the Government is added, basically through public research centres and universities, it becomes clear that resources obtained by the public sector are superior to those of the private sector.

As regards the application of funds, with relation to their origin, the situation in Greece also differs to that of the other countries with high dependency on the public sector.

Figure 3-7: Expenditure on R&D by sector (%)

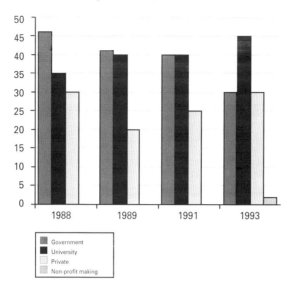

Source: GSRT, 1996

Table 3-23: Origin and Use of Funds Intended for R&D in Greece (Millions of Drachmae - 1993)

Origin of funds	Use of Funds					
	Government	University	Non-profit institutions	Companies	Total	%
Government	21,562.0	6,426.0	154.0	1,878.0	30,030.0	29.7
University	- -	17,440.0	- -	- -	17,744.0	17.6
Non-profit institutions	- -	- -	152.0	- -	152.0	0.4
Companies	405.0	1,551.0	54.0	19,063.0	21,073.0	20.8
Foreign	10,157.0	12,697.0	211.0	8,798.0	31,863.0	31.5
Total	32,124.0	40,841.0	571.0	29,739.0	100,862.0	100.0

Source: OECD, 1995

In Table 3-23, within that financing included as "Government", we have taken into account that the immense majority of these funds in one way or another originate abroad through funds provided by the European Commission. This paragraph deals with European participation in the financing of the R&D activities, a theme not dealt with here previously. As indicated beforehand, the EPET II programme came within the Community R&D Support Framework. Likewise, the majority of public capital employed in scientific research has its origin in EU funds. This great imbalance of EU financing as compared to public financing can have serious negative consequences in the near future according to the GSRT, if the ratio does not change by increasing government expenditure. This is the only means by which appropriate use may be made in the long run of experiences, knowledge and infrastructure that has resulted from participation in the various support programmes of the European Union.

With respect to personnel involved in R&D related areas, the figure has increased in the last decade in line with the other indicators of the S&T system.

Figure 3-8: Personnel Related to R&D

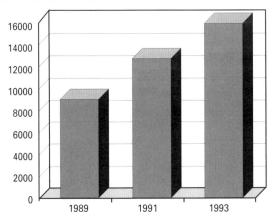

Source: GSRT, 1996.

It can be deduced from Figure 3-8 that the number of personnel related to R&D activities doubled in the period 1989-1993, a rate similar to the rate of increase of R&D expenditure in constant prices. The distribution per sector is shown in Table 3-23.

Table 3-23: Distribution of R&D Related Personnel (1993)

Sector	Administration	University	Non-profit Institutions	Companies
Scientists	1,905	4,773	34	1,338
Technical	971	1,350	12	941
Other	1,952	644	28	652
Total	4,828	6,767	74	2,931
%	33	46.3	0.7	20

Source: OECD, 1995

The imbalance between the number of researchers in the public sector, including universities, and the number of researchers in the private sector is large. In other EU countries the imbalance is not so great. Only in Portugal, with a 18%, on the total, can similar rates be found.

REFERENCES

Common statement on research and technology from the secretary generals during the period 1982-1995.

Economist Intelligence Unit, Country Profile: Greece. 1995-1996.

European Monetary Fund (1996) Perspectives of the World Economy.

Eurostat, Europa en Cifras (4° Ed.)

General Secretariat for Research and Technology (1996), ACTIVITIES "A Window in the Future".

General Secretariat for Research and Technology (1996). Bulletin Research and Tecnology in Greece.

General Secretariat for Research and Technology Bulletin Research and Tecnology in Greece, January, 1997.

General Secretariat for Research and Technology, Bulletin Research and Technology in Greece, August, 1996.

Ministry for Development, General Secretariat for Research and Technology (1996), A Guide to Bilateral Cooperation..

Ministry for Development, General Secretariat for Research and Technology (1996), Greek Research and Technology Network. GR-NET..

Ministry for Development, General Secretariat for Research and Technology (1994), Operational Programme for Research and Technology.

OECD (1986), Reviews of National Science and Technology Policy: Greece: 1984.

OECD (1991), Report on Scientific and Technological Policies, Balances and Previsions

OECD (1992) Technology and the Economy: Key Relationships.

OECD (1994), Science and Technology Policy. Review and Outlook.

OECD (1995), Statistiques de Base de la Science et de la Technologie.

OECD (1996) OECD Economic Studies: Greece 1996.

Office of Official Publications of the European Community (1995), General Report on the Activity of the European Union

Other information made available by the General Secretariat for Research and Technology

ITALY

Chapter 4

ITALY

INTRODUCTION

The Italian S&T system, within the context of the four countries being analysed, is the most developed. Comparison with the other EU countries still places it in an acceptable position. Reference can be made to competitive research in wide sectors of activity and it will be seen that the system is among the ten best, world-wide.

Its structures, both in terms of organisation and financing, are comparable to those of developed countries, while the main indicators show that a substantial increase in effort has taken place in the last 20 years. Regarding R&D, Italy is at the level of the EU average and despite being slightly stagnant in recent years, the perspectives of growth are encouraging.

The nature of the Italian economy results in substantial differences between the Italian S&T system and those of the other countries examined here. The productive sector is strongly dominated by exports from certain sectors, especially those of electrical applications, cars and textiles. The maintenance of acceptable levels of competitiveness demands wide R&D activity in order to develop new innovative processes, although this is not the case in Italy. Innovative processes are sustained by the acquisition of external technology (which is later being adapted to concrete needs) rather than by effective research. On the other hand, the presence of large multinationals has to a certain extent eclipsed the work of SMEs, especially in topics related to scientific research.

Another peculiarity of the Italian economic system is the problem of Mezzogiorno, located in the south of Italy. While the regions located to the north are prosperous and the standard of living in no way compares unfavourably with that of the more developed regions of the European Union, regions in the south are much less developed. The gravity of this problem has lessened to such an extent that at present, the gaps existing with regard to research are not greater than those of other countries with lower regional imbalance. This is confirmed by examination of the levels of participation by regions in the European Union Framework Programmes.

In general, the system is highly consolidated, given that its institutions, especially the National Research Council (*CNR*), and the supporting legal framework have been carrying out their activities for many years, and always along the same lines of action.

INSTITUTIONAL FRAMEWORK OF S&T POLICY IN ITALY

Figure 4-1: Structure of the Italian S&T System

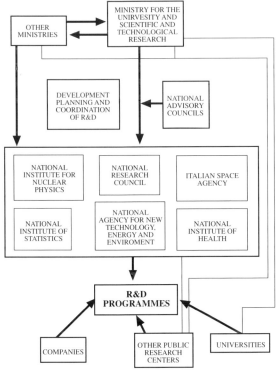

Ministry for the University and Scientific and Technological Research

The Ministry for the University and Scientific and Technological Research (*Ministero dell'Universita e della Ricerca Scientifica e Tecnologica - MURST*) is the main body responsible for the S&T system in Italy. Its main functions are the promotion of scientific and technological research and the development of universities and associated institutions. The Ministry defines the scientific policy that must be implemented in the country, as well as establishing the most appropriate mechanisms for its development. In addition, it coordinates Italian participation in international, bilateral or multilateral projects, both within the European Union and beyond, incorporating these projects to the structure of national scientific projects.

MURST assigns economic resources to the development of scientific research, negotiating agreed private funds dedicated to the programmes and orients training policies for researchers at all levels, proposing measures for the development of employment and facilitating researcher mobility in both research and production.

The Ministry is structured in diverse departments and services, each one of which carries out specific functions and takes charge of defined areas within Ministry activities. These departments are:

- Department of Planning and General Coordination.
- Department of University Education.
- Department of Scientific and Technological Research
- Department of International Relations.

These departments are responsible for the development of R&D policy in fields related to research planning and coordination, the university system, applied research, research based on objectives, international relationships, and legal aspects related to science and technology.

In order to carry out this series of activities, *MURST* has created specific organisations to provide the necessary support for the correct evolution of the S&T system in accordance with national interests and in coordination with the actions carried out within the framework of the European Union:

- National Council for Science and Technology.
- National University Council
- National Geophysics Council
- National Council of Astronomic Research

In addition to their function as advisors to the Ministry for the University and Scientific and Technological Research, these centres also serve as links among scientific groups, universities and public research centres. A short description of each is given below.

National Council for Science and Technology

The National Council for Science and Technology (*Consiglio Nazionale della Scienza e della Tecnologia - CNST*), the majority of whose members are elected, acts as an advisory organisation to *MURST*, the Presidency of the Government and the Council of Ministers.

The Council advises and makes recommendations on policies to be developed through studies carried out by the universities and other agencies involved in aspects related to S&T. With respect to annual and pluri-annual planning, these studies are fundamental when elaborating the policies to be implemented; priorities, resources to be assigned to each area, Italian participation in international cooperation programmes, and development of the different sectors of the S&T system in order to reach the proposed objectives.

National University Council

The National University Council (*Consiglio Universitario Nazionale - CUN*) is an elected body representing the Italian universities. It advises the Minister for the University and Scientific and Technological Research on the following:

- Coordination of university activities.
- Recruitment of university professors and scientists in accordance with the needs of these institutions and the relevance of research projects to be developed.
- Definition and updating of the national educational organisation and the triennial university development plan.

National Geophysics Council

The National Geophysics Council (*Consiglio Nazionale Geofisico - CNG*) coordinates geophysical and vulcanologic research developed by the Observatory of Vesuvio and other public agencies or organisations involved in this area. Apart from its activities at institutional level, the Council also carries out seismology studies of the national territory in collaboration with university institutions and other public research centres. Activities are supervised directly by the Ministry for University and Scientific and Technological Research.
 Other activities are:

- To present proposals for research projects to the Minister in the areas of vulcanology and seismology.
- To evaluate and follow up relations between national and international agencies developing activities related to geophysics.
- To establish new geophysical observatories in the national territory.
- To propose and evaluate Italian participation in international projects and assign funds accordingly.
- To improve the infrastructure of Italian geophysical observatories by means of appropriate financing.

National Council for Astronomic Research

The National Council for Astronomic Research (*Consiglio Nazionale per le Ricerche Astronomiche - CNRA*) coordinates the research activities carried out by the network of Italian astronomic observatories. It also coordinates activities related to astrophysics in public research centres working in this area. As these projects are of a considerable magnitude, great importance has been given to coordination with international activities. The Council advises on the assignment of funds to the national network of observatories, astrophysics programmes being carried out in Italy and on the recruitment of scientific personnel for the development of new initiatives.

A new law is expected in September 1997 which will result in the reform of the insitutions described above and some of the research centres to be described below. Although it is probable their general functions will remain unchanged the possiblity also exists that some functions may vary.

PUBLIC RESEARCH CENTRES

National Research Council

The National Research Council (*Consiglio Nazionale delle Ricerche - CNR*) is not only the most important institution of the Italian S&T system but also serves as the main link between the various decision making bodies with respect to the Italian S&T system (mainly Ministries) and the sectors of Italian society which are the true executioners of these policies (basically companies, universities and public research centres).

Set up in 1923, the *CNR* has come under the direct supervision of the Minister for the University and Scientific and Technological Research since 1989.

Its initial purpose was the promotion, coordination and regulation of scientific research in Italy, even carrying out activities which today would be considered of ministerial responsibility. Throughout the years, its responsibilities have evolved in line with the changes in scientific-technological relations, links between institutions and the Italian socio-economic background (creation of the Ministry for the University and Scientific and Technological Research, the university law, international cooperation programmes).

The most outstanding activities of the *CNR* at present are:

- Realisation of studies and consultancy to legislative bodies with respect to research and technology.
- Negotiation of research contracts
- Financing of R&D activities
- Granting of scholarships to national and foreign students
- Research activity through its research bodies.

Although the *CNR* depends directly on the *MURST*, it has its own objectives as an institution. These objectives are laid down by the *CNR* President, the Executive Director, the Presidential Council and the Administrative Committee. It may also avail of the aid of consultative boards (the National Advisory Committees and the Plenary Assembly).

In 1994, the budget was approximately 1,708 billion lira, which provides an idea of the size of this organisation. In the same year, the number of staff was 7,000, of which 4,000 were researchers working throughout the country in the network of public research centres. Another 2,000 reseachers are employed on a contract basis or as grant-holders.

Figure 4-2: Evolution of State Funding to the *CNR* (1989-1994) *

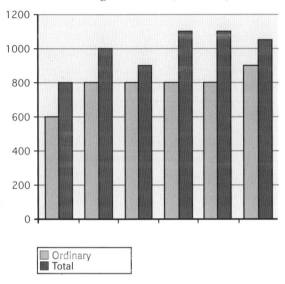

Quantities in Billions of Lira.

Source: CNR - Balance 1994

Institutes

The Institutes are permanent operative organisations whose objectives are in direct correlation with those of the *CNR*. Personnel and teams necessary for the development of activities are directly provided by the *CNR*, which also covers all running costs.

Study Centres

Such Centres are set up in universities, scientific organisations, public administration and in some private organisations. Their mission is the development of particular studies and advanced research. The *CNR* and other collaborating bodies provide the personnel and teams necessary for the development of activities.

Research Groups

These Groups are temporary organisations set up for a maximum period of 5 years, although on occasion their duration may be longer. They develop organisational tasks with the purpose of coordinating the creation of new research centres or new organisations made up of diverse groups of researchers.

---

* Lira. Average exchange rate in 1994. 1.612 lira=1$. 1.909 lira=1 ECU. 995.995 lira = 1 Deutschmark

Research Bodies

Institutional scientific activity is directly developed by operative research bodies made up of 163 Institutes, 114 Study Centres and 18 Research Groups.

The Institutes are research organisations, depending on and collaborating with the *CNR*, while the other entities carry out activities in coordination and collaboration with other organisations responsible for scientific research.

The activities and operations of these bodies are controlled and supervised by 15 National Advisory Committees, which the former may also consult. The Committees are organised according to the areas in which the *CNR* is involved:

- Mathematical Science
- Physical Science
- Chemical Science
- Biological and Medical Science
- Geologic and Mineral Science
- Agricultural Science
- Engineering and Architecture
- History, Philosophy and Psychology
- Juridical and Political Science
- Economic, Social and Statistical science
- Technological Research and Innovation Science
- Information Sciences and Technology
- Science and Technology of the Environment and Habitat
- Biotechnology and Molecular Biology
- Sciences and Technology for Artistic Heritage

Research Areas

Research Areas could be described as cooperation networks integrating *CNR* institutions carrying out activities in the same regions. They are are located in:

- Torino
- Pisa
- Naples
- Sassari
- Cosenza
- Milan
- Bologna
- Bari
- Potenza
- Padova
- Florence
- Palermo
- Cagliari
- Genova
- Rome
- Catania
- Lecce

Determinant Projects

In the last decade, the *CNR* established a series of 5 year programmes. These so called Determinant Projects are carried out in accordance with the main national interests and current socio-economic needs. All groups belonging to the Italian scientific network (public and private organisations) participate in such projects.

Programmes are diverse, covering areas from advanced technology to environment, from human health to agricultural or food resources. The results of these projects have been published in more than 6,300 scientific articles in the Italian and international press.

To date, two complete cycles of Determinant Projects have been completed, and the "third generation" is also reaching its end. The 18 projects are:

- Telecommunications
- Robotics
- Optic-Electronic Technologies
- Fine Chemicals
- New Specialised Materials for Advanced Technologies
- Superconductive and Cryogenic Technologies
- Internationalisation Enterprise
- Biotechnology and Bioinstrumentation
- Construction
- Information Systems and Parallel Computing.
- Genetic engineering
- Aging
- Prevention and Control of Disease Factors
- Clinical Applications of Oncological Research
- Advanced Research for Innovation in Agricultural Systems
- Transport 2
- Organisation and Operations of Public Administration
- Materials and Devices for Solid State Electronics (MADESS)

Strategic Projects

The *CNR* as part of its task of defining the needs of the socio-economic context at the technical level regarding R&D in Italy, draws up a list every two years of the more important projects. These projects are then labelled "Strategic Projects" and are considered of national and international importance. The first list contained the following projects:

- Information Technologies
- Environment and Territory
- Advanced Technologies in Biology
- Chemical and Physical Aspects of Biological Systems
- Advanced Technologies
- Infrastructures and Services
- Cultural Themes
- Fusion including the RFX Tritium and Microwave
- Geographical, Naturalistic, Physical, Anthropic Studies and Medico-Physiological Surveys in the Himalayas and in Karakorum.

The reason for distinguishing between the two types of project, Determinant and Specific, is that the former require the participation of structures or organisations outside that of the *CNR*, while the latter are carried out in their entirety by *CNR* personnel and institutions.

Programme design must be compatible with the organisation itself, while actual development demands proven research capacity.

International Activity

Since the date of its creation, the *CNR* has played an important role in the scientific relationship of Italy with other countries. At present, it has 76 collaboration agreements with other non-governmental foreign institutes, and participates in international projects promoting research in collaboration with other organisations and favouring the exchange of research personnel. The number of agreements signed with homologous institutions in other countries at present is 33. The participation of *CNR* in the development of projects within the EU Framework Programme, such as nuclear fusion, environment, electronics and computer science, should also be highlighted. The *CNR* also develops projects in collaboration with Italian universities and in conjunction with other foreign institutions.

Main CNR International Projects

Some of the main international projects of the *CNR* are the following ones:

- Spallation Neutron Source (SNS) at the Rutherford Appleton Laboratory (RAL) at the SERC
- Ocean Drilling Programme (ODP)
- European Science Foundation (ESF)
- Human Frontiers Science Programme (HFSP)
- European Research Consortium for Informatics and Applied Mathematics. (ERCIM)
- European Atomic and Molecular Computing Centre (CECAM)
- International Computer Science Institute of Berkley-CA (ICSI)
- European Very Long Baseline Interferometry (VLBI)
- Large European Solar Telescope (LEST)
- Programmes in the Arctic and the Antarctic

As part of its international activity, the *CNR* has recently opened the Office for Promotion and Relations with the European Commission in Brussels.

The Central Library

The Central Library, founded in 1927, is the largest Italian scientific library containing an immense number of publications on Italian and international research.

It has a total of more than 600,000 volumes and 10,000 periodicals which occupy an area of more than 17 Km. of bookcases, with 5 reading rooms open to the public.

The *CNR* Central Library coordinates all the scientific libraries of its institutes which are spread throughout the country and organises professional training courses for the staff. Documentation services are also provided to other institutions and organisations involved in scientific research. It is responsible for the classification of the bibliographical heritage of the *CNR* and promotes the exchange of information with external institutions.

Database

The *CNR* produces a great quantity of documentation on science related topics, which must also be made available to the scientific community as a whole. Special effort has been made in last decade when classifying that information relevant to the activities of the S&T system and especially the work developed in public research centres. This is presently available through a database, and is the first time that research groups have efficient and flexible access to such a wide volume of information.

The consequences of the installation of this database will strengthen interaction among the different work groups involved in the development of parallel activities. Their main objective is that of making information accessible so that any researcher can have access to the experience accumulated by other scientists in topics which are of interest. References on more than 5,000 projects are available in addition to data on the entities responsible for carrying out specific plans of scientific research. Archives also exist containing detailed results of programmes and projects developed in the previous 5 years.

The work of the *CNR* in the Italian S&T system is of vital importance. Not only is it the "institute of institutes" coordinating the main part of public research in Italy, but given the length of time it has been in operation, it is highly consolidated within the Italian social fabric.

Italian National Entity for New Technology, Energy and the Environment

The Italian National Agency for New Technology, Energy and the Environment (*Ente Nazionale per le Nuove Tecnologie, l'Energia e l'Ambiente - ENEA*) is a government agency controlled by the Ministry for Industry. Its two fundamental tasks are that of driving the research in the areas of new technology, energy and environment and disseminating the conclusions and results at national level. Research lines are aimed at the development of Italian industry, but in accordance with the legally established environmental requirements. Through the National Agency for the Protection of the Environment, it is also responsible for the control of nuclear installations and pursuit of labour safety norms in companies.

The general lines of research developed in this agency are as follows:

- Research and development of new technologies and equipment, and transfer of results to the industrial and agricultural sectors
- Development of a specific market in technology, equipment and components designed for the appropriate exploitation of renewable energy and the reduction of energy expense in general. Design, construction and follow up of new energy plants in Italy.
- Research and development of new nuclear reactors, improving safety, at the same time dismantling old energy producing power stations, especially nuclear plants.
- Development of energy processes based on nuclear fusion in close collaboration with international scientific groups.
- Monitoring and supervision of activities susceptible to affecting the environment. Development and installation of new technologies with low environmental impact.

The number of staff is 5,000. The organisational structure of the agency is based on nine institutes, spread throughout the entire country.

Specific programmes, whose general lines have already been mentioned, are in accordance with central administration policies. However, the activities of the Agency are not only limited to the development of large programmes or areas of action. There are countless initiatives which are dedicated to supporting Italian enterprise, especially smaller companies. In order to stimulate innovation, the *ENEA* promotes the installation of new technologies, the result of research processes carried out, in production. The *ENEA* is at present involved in projects of collaboration with small local companies to develop and install new technological systems, as well as aiding larger companies to improve competitiveness.

Figure 4-3: *ENEA* Budget in the period 1986-1994

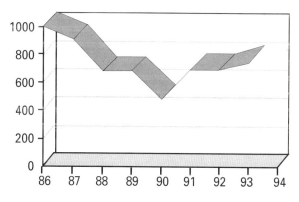

Quantities in Billions of Lira

Source: ENEA

In addition to its own research, the *ENEA* also carries out a series of other activities in order to achieve its objectives; environmental audits in companies, reports on the installation of new technologies, establishment of the legal aspects and frameworks that enterprise must contemplate when involved in the *ENEA* areas of responsibility.

The evolution of the budget of the *ENEA* can be seen in Figure 4-3.

Italian Space Agency

Italy's involvement in space projects can be traced back to more than 30 years ago, but it was not until 1979, that a series of actions were initiated in order to guide the different programmes, which given their previous lack of coordination were to a certain extent incoherent. Due to the impulse given to space programmes world-wide, it was necessary to create a specific agency unifying the policies that should be followed. Therefore in 1988, the Italian Space Agency (*Agenzia Spaziale Italiana - ASI*) was set up with the purpose of rationalising activities.

The *ASI* is responsible for the S&T planning of programmes in the aerospace sector and improving the competitiveness of the Italian aerospace industry.

It develops the National Aerospace Plan, originally managed by the *CNR*, and coordinates actions with those of other homologous institutions world-wide, in particular with those of the European Union. It also coordinates Italian participation in the European Space Agency (ESA).

Figure 4-4: *ASI* **Budget in the period 1986-1994**

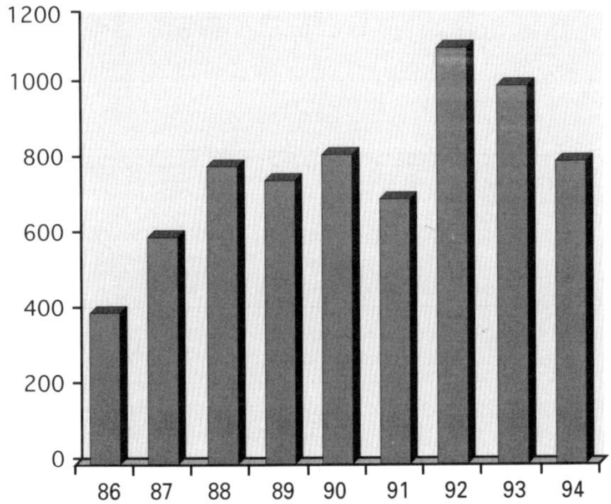

Quantities in Billions of Lira.

Source: ASI.

Given the nature of aerospace projects (high levels of investment and the participation of a large number of specialists and scientists), coordination among actors is vital. Most initiatives are developed through universities or contracts with private companies.

In the period 1990-1994, the Government approved an annual budget of approximately 1,5 billion dollars to *ASI* activities. The evolution of funds can be observed in Figure 4-4. Public administration policy throughout this period was that of maintaining Italian participation stable in ESA and the national aerospace programmes. In terms of participation in ESA, Italy presently contributes 18% of the total budget. The collaboration of *ASI* with NASA is of the same magnitude in economic terms as that of participation in the European Space Agency.

National Institute for Nuclear Physics

The National Institute for Nuclear Physics (*Istituto Nazionale di Fisca Nucleare-INFN*) is the public organisation responsible for research in areas related to nuclear physics. Certain responsibilities are shared with the *ENEA*, although *INFN* concentrates only on this area. Activities are developed through 19 sections located in the same number of Italian universities, 7 operative groups, 4 laboratories and other public institutions. The sections are totally independent with respect to both administration and actions, depending only on the *INFN* and not on the university. The link with the latter is through specific agreements.

Figure 4-5: *INFN* **BUDGET 1986-1994**

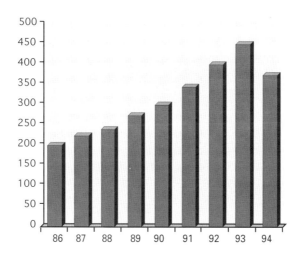

Quantities in Billions of Lira.

Source: INFN

R&D plans are developed during 5 year periods and are designed entirely by the *INFN*, which then presents them to the *MURST* for approval. Programmes to be carried out in the period 1994-1999 are presently in force.

INFN Laboratories and Sections are the main instruments for the development of specific programmes. The Laboratories are integrated into the facilities of main public research centres throughout the country, thus ensuring that access is not limited only to *INFN* scientists. The Sections work closely with the universities. At present, more than 2,000 people linked to the university environment collaborate in *INFN* activities. The budget of this institute can be observed in Figure 4-5.

One of the most important facets of the Institute's activities is international collaboration with other foreign homologous organisations. Almost 60% of the total budget of the *INFN* is dedicated to programmes involving international research centres in one way or another. This organisation coordinates Italian participation in CERN, providing this centre with more than 600 scientists.

An intergovernmental agreement between Italy and the United States of America, resulted in coordination by the *INFN* of Italian participation in North American research centres related to nuclear energy. Within this framework, *INFN* working groups were able to develop their activities in: the Fermi Lab in Chicago, Linear Accelerator of Stanford; National Laboratory of Brookhaven, etc. Approximately fifteen North American work groups collaborated in the activities of the Italian laboratory in Great Sasso.

National Statistics Institute

The National Statistics Institute (*ISTAT*) is an organisation of the National Statistics System, created in 1988 and answering directly to the Prime Minister. This Institute is also considered a research centre.

Among the institutes making up the national network of statistics offices, those to be highlighted are those corresponding to: Government, regional administrations, Chambers of Commerce, industries, agricultural, etc.

Statistical work is based fundamentally on related national programmes, which are reformulated every three years. The programmes undergo examination by the National Committee for the Security of Statistical Data which presents its report to the Government. If this is satisfactory, the President of the Republic approves.

ISTAT is the main body responsible for the National Statistics Programme, which basically consists in: the development of censuses and statistical studies; coordinating, directing and providing technical support to the offices making up the National Statistics Network. *ISTAT* is also involved in the establishment of procedures and nomenclature that must be used in statistical analysis, and of analysing and researching the reasons for the results of relative studies on phenomena of national interest.

The Institute carries out its work in collaboration with other national agencies or private companies with which it has cooperation agreements.

Of the total of 2,800 personnel working in the National Statistics Institute, more than 400 are researchers.

National Health Institute

The National Health Institute (*Istituto Superiore di Sanita - ISS*) is the main Italian institution for technological and scientific research in human health. Its foundation dates back to 1934 and it was in 1978 when it began to develop its work within the National Health Service as its main scientific and technical support.

Although it comes under the supervision of the Minister of Health, the *ISS* has its own structures and a high level of autonomy, especially with respect to scientific research.

Its mission is that of improving human health through studies and analysis of scientific character in the context of Italian society. Main lines of research are concerned with the following medical aspects:

- Infectious and non-infectious pathologies
- Environment and its influence on human health
- Food
- Drugs

Research results are disseminated through conferences and seminars in the medical community, improving training formation and keeping it up to date on the latest advances in health.

It collaborates with the Ministry of Health in the preparation of the National Health Programmes and proposes health regulations to be established in Italy.

The activities of the *ISS* are not exclusively limited to those directly related to the Ministry of Health. Another series of work financed by different entities or public agencies is also carried out. Collaboration with the European Union and with the World Health Organisation (WHO) is very significant. In collaboration with the latter, the National Institute for Health develops specific programmes relative to:

- Influenza surveillance
- AIDS
- Surveillance of infections
- Research and training on veterinary aspects
- General training of personnel of the National Health System

The Institute consists of 20 laboratories and 8 technical assistance centres, while it employs almost 1,500 staff. It also has an extensive library with more than 150,000 volumes and 3,500 periodicals.

Other Public Research Centres

Other public research centres also exist in Italy which are smaller in size and specialise in very expert areas of R&D. Usually such centres work in coordination with, or provide a service to other larger public institutes.

The National Institute for Nutrition, the body responsible for biological studies of food, works in cooperation with the *ISS*, while the Institute of Social Medicine (*Istituto italiano di medicina sociale - IIMS*) centres on medical and social problems of health at work, including prevention and assistance. The Higher Institute for the Prevention of Accidents and Labour Safety works in collaboration with the latter. The Higher Institute for the Prevention of Accidents and Safety at Work (*Istituto superiore per la prevenzione e la sicurezza del lavoro - ISPESL*) covers the four areas of research of factory production, environmental impact, technologies related to labour safety and hygiene.

With respect to environment, the National Geophysical Institute (*Istituto Nazionale di Geofisica - ING*) and the Experimental Geophysical Observatory (*Osservatorio geofisico sperimentale - OGS*) conduct research on the study and prevention of earthquakes and other natural catastrophes, as well as studies on seismology, geomagnetism, etc.

In the area of production, the activities of the National Electrotechnical Institute Galileo Ferraris (*Istituto Elettrotecnico Nazionale "Galileo Ferraris"*) is particularly strong, carrying out permanent studies on metrology, production techniques, processing of new materials and systems engineering. Another agency operating in the same sector is the National Institute of Studies and Experiments in Naval Architecture (*Istituto nazionale per studi de esperienze di architettura navale- INSEAN*), involved in the development of projects related to naval design and hydrodynamics.

An aspect of special relevance are institutes focusing on the area of social sciences and economics such as:

- *ISPE*: Institute for the Study of Economic Planning (*Istituto di studi per la programmazione economica*)
- *ISCO*: Institute for the Study of Economic Trends (*Istituto nazionale per lo studio della congiuntura*)
- *ISFOL*: Institute for the Development of Professional Training for Workers. (I*stituto per lo sviluppo della formazione professionale dei laboratori*)

The *ISPE* develops specific programmes relating to the supervision of public finances, analysis of public structures, effects of fiscal measures, etc. The *ISCO* produces reports on the Italian economy in the short term, and is involved in the analysis of the capacity of the Italian productive sector and its influence on the national socio-economic context (prices index, product demand, etc.). The *ISFOL* plans professional training for workers and analyses changes in the structure of employment. All three institutions combine research with the provision of services to public organisations or companies.

The "Francesco Severi" Higher Institute of Mathematics and the National Optics Institute (*INO*) are noteworthy among the smaller institutes concerned with particular areas of science. Although funding is low, the activities carried out here are essential given their high degree of specialisation. The former is concerned with the training of researchers in the area of exact sciences while the latter develops studies related to optics. It also provides assessment and consultancy to enterprise and other national scientific institutions.

Overall resources dedicated to the maintenance of these centres is greater than 150 billion liras. Of the total staff of 1,000 professionals approximately half are researchers.

Other institutions, of a non-public nature, are permanently involved in R&D activities. Of these the most outstanding is the *ENEL* (National Electricity Board) which does substantial research. Main projects are concerned with: the study of improvements in energy to be incorporated into the production sector; the improvement of the national electricity network; the improvement of the process of electric power production in order to reduce environmental impact, etc. The research carried out here is supervised by the Central Research Division, also responsible for the management of the five *ENEL* research centres:

- Automation (*CRA*)
- Electricity (*CREL*)
- Hydraulics and Structures
- Nuclear and Thermal (*CRTN*)
- Coal (*CRC*)

The number of full-time researchers employed in *ENEL* is 300, to which can be added personnel from other companies, collaborating or carrying out research through agreements.

R&D PROMOTED BY MINISTRIES.

The main actions of the Ministries in the promotion, planning and coordination of scientific research will now be examined. In comparison with public research centres, whose objectives tend to be broader in scope, research carried out by the Ministries is more directly concerned with the problems of the sector in question; they try to provide solutions to the fundamental needs of Italian society. In spite of this, the importance of basic research (an area usually assigned lesser importance in the actions of the Ministries) should not be forgotten, given that it plays an essential role in the continuity of the scientific process. The main activities of each Ministry and actions carried out in recent years are outlined below.

Ministry of Agriculture

The Ministry of Agriculture has 23 research institutes, the activities of which are related to national agricultural policy, and which in turn constitute 91 operative central sections and 53 local sections nation-wide.

In line with the projects developed, these Institutes can be divided into three groups:

- Institutes concerned with specific disciplines (protection of the earth, agricultural systematizing, vegetable nutrition, automation of farms, agricultural zoology, etc...)
- Institutes responsible for the cultivation of new plants, with the objective of increasing crop productivity and resistance to external aggressions.
- Institutes which develop new technological processes for agriculture.

Areas of research are in line with National Agricultural Plans, of which the first two phases have already finished. The most important of these areas are related to:

- Improvement of the quality of agricultural products.
- Development of new seed varieties.
- Control and orientation of production towards products where greater demand is expected.
- Development of biotechnology in the agricultural sector

Collaboration projects to improve agricultural practices are also carried out with other institutes within the framework of the National Agricultural Plan.

Ministry of Culture

The research activities of the Ministry of Culture are concerned not only with humanities; the experimental character of some of its activities is of great importance. Given the amplitude of the Italian artistic heritage, the development of research on the restoration of artistic and archaeological monuments is essential. There are a high number of Departments situated throughout the entire nation, carrying out geological tests to detect historical remains, on which all types of physical and chemical tests are carried out. The Departments can make use of public research centres and institutes and other state organisations such as the Central Restoration Laboratory.

Scientific research in the Ministry of Culture is also to be found in the maintenance of the national historical archives, and is conducted by the Department of Archives with the aid of archivists' institutes and schools, paleographers, etc. which collaborate in work characteristic of this Department.

The total of personnel related to research in this Ministry is 1,500, half of whom are researchers.

Ministry of Industry

The Ministry of Industry supervises the activities of the public research centres *ENEA* and *ENEL*, whose R&D activities have been previously described.

The Ministry also controls the activities of the experimentation stations which are an interface between research carried out in public centres (in many cases, basic research) and production (applied research). Taking into account normal enterprise activities, the work carried out in the experimentation stations is concerned primarily with: the optimisation of production processes; utilisation of industrial residuals; recycling; analysis and testing; certification of new products and environmental protection. There are eight such centres, specialising in:

- Oil and Fats
- Cellulose and Textile Fibres
- Silk
- Fuel
- Glass
- Food Conservation
- Leather Tanning
- Essences

Ministry of Health

The research activities of the Ministry of Health centre on projects carried out by the national central institutes previously described (National Health Institute and the Higher Institute for Labour Safety). The main lines of research are:

- Epidemiology
- Evaluation and provision of services
- Economy
- Management
- Training of health personnel
- Relations between citizens and the National Health System
- Health education

Ministry of Special Measures for the Area of Mezzogiorno

A particularity of the Italian State is the difference between the regions of the south and those of the north. Although such differences are various, that of greatest concern to the S&T system are those of an economic nature. While the regions of the north are highly developed and prosperous (in many cases the standard of living is higher than the EU average), the south is poor, with low levels of development and investment by industry. Although such differences (a rich northern area in contrast with lower economic resources in the south) also exist in other countries, Spain or Portugal for example, the case of Italy is intrinsically different. In the other two countries the differences, in spite of being significant, are not excessive, while in Italy this is so. The economic power of a region or country influences decisively on the level of R&D investment. It is sufficient to compare any R&D expenditure with the GDP or any other indicator of economic development to realise that development is parallel.

The Italian State set up a specific ministry to promote research in the regions of the south, the Ministry of Special Measures for the Area of the Mezzogiorno. Although the Ministry no longer exists, and its powers transferred to other ministries, brief mention of its main activities in the promotion of R&D is justified given the importance of the initiatives carried out.

Such measures included the development of R&D in enterprise in these regions. The objective was the integration of companies and research centres located in the south with those of the rest of the country in order to achieve long term economic cohesion.

Decree 218/78 constituted the legal support of these specific actions, laying down funding for the research centres, while Decree 64/86 established the financing of scientific projects. Initially, the channel set up to obtain such economic resources was through concrete agreements of industry with the Government or through individual applications.

Management control mechanisms of the projects developed in the regions of the south depended directly on the Ministry of Special Measures as well as on the Agency for the Development of the South of Italy.

The most recent research support measure in the regions of the Mezzogiorno will be promoted (mid-August 1997) by the Committee of Economic Planning *(Comitato interministeriale per la programmazione economica - CIPE)*, and through which the sum of 500 billion lira will be made available for investment, serving as a complement to the research activities already being developed in the area. The initiative is considered vital by the *CIPE* as it will constitute an indispensable instrument for the improvement of SME competitiveness and employment.

New mechanisms for the solicitation of projects have also been designed in the framework of this initiative, simplifying the bureaucratic process and reducing the waiting periods.

Ministry of Defense

In the context of the countries being dealt with in this book, Italy's military industry is the most developed (Spain is also specially competitive in this aspect, and in terms of turnover is not far behind Italy, although its technological level is significantly lower). The objective of this industry is to maintain military material up to date and achieve export contracts with third countries. It should not be forgotten that the sophisticated military materials undergoes significant renovations as it is prone to becoming quickly obsolete. This is not so in the case of less complicated material whose life cycle is longer. In any case, scientific research is a key element in the overall process and is worth commenting on.

Since the Second World War the military industry and the advances which it has undergone have provided important technological progresses to society as a whole. The aerospace industry for example has been basically stimulated on advances taking place in the area of military aviation. The need to carry out projects of high investment which involve a large number of companies has facilitated the creation of communication links among the different elements making up the production sector. That is to say, the military industry impels national industry overall to stimulate its competitiveness when having to fight in markets dominated by more developed countries.

R&D in this sector has, for obvious reasons, a strong state control and it is supervised by high ranking military staff. These projects are carried out by public research centres on demand for a noteworthy number of companies together, which occasionally develop projects under their own initiatives.

In Italy, the actual number of researchers equivalent to full dedication involved in military R&D projects is over 900, of which about 300 are researchers and the remainder, specialised technicians.

Ministry of Transport

Research in the area of transport is carried out in line with the General Plan for Transport, which establishes the level of financing to be granted to R&D activities carried out by the institutes involved.

The organisation, State Railways, is strongly linked to R&D activities through: specific projects for the improvement of the railways; the design of prototypes; and the improvement and installation of high speed lines. Research is financed through contracts established with industry, which include funding for research in their budgets.

Although investment in this type of infrastructure is high (in the 80's it was more than 290 billion lira in rolling stock alone), on many occasions R&D investment does not consist of the development of own technology as a solution to specific problems, but rather the acquisition of foreign technology, which is then adapted to the specific needs. This policy solves short term problems, but to a certain extent it compromises the future competitiveness of Italian industry. Greater concentration of effort is therefore necessary in the realisation of own projects.

The most important organisation in charge of transport research is the Experimental Institute, part of the State Railways and founded in 1905 as a scientific and technical body. Its activities are directly related to; electronics, electronic engineering, physics, chemistry, metallic and non-metallic construction materials, geology, spectrography, refrigeration and environment. The environmental laboratories consist of independent mobile units which carry out field studies on the environmental impact of railroad infrastructure and traffic. They also analyse the best location for new railroads.

Ministry of Mail and Telecommunications Services

Through its scientific-technical branch, the Higher Institute of Postal Services and Telecommunications (*ISPT*) carries out specific research projects with the Ministry of Mail directly related to the improvement of the postal service and telecommunications. Most of these projects are aimed at the installation of new organizational procedures. This Institute is also involved in the area of telecommunications, an area which is acquiring greater importance. Given the high volume of capital invested in the installation of new communications systems, the financing of R&D activities in this field is considerable. The *ISPT* has cooperation agreements with homologous foreign institutions in the following:

- Telecommunications
- Telematics
- Postal Services
- Information Technology

The number of personnel related to scientific research is 800. Important effort is being made in the area of specialist training by means of cooperation agreements with the Higher School of Telecommunications.

Activities of Local Administration

The role of the local authorities in Italy is strongly linked to scientific research. Through regional government and city council initiatives, specific plans are laid down for the financing of R&D activities in those sectors of greater interest to the region. Most of these plans or programmes are carried out in agreement with local institutions, such as universities, public research centres or groups of companies. Another common form of research development, cooperation and dissemination of results is through the science and technology parks spread throughout the nation, and whose management corresponds to the local authorities. The most important are Technocity in Piamonte, the Technopolis in Apulia or the Consortium for the Research Area of Trieste.

The main problem of this type of investment is that adequate coordination with the specific national plans for R&D does not exist. Being isolated investments which respond to short term needs, the risk exists that the common aim established at the national level by the overall set of actions is lost. Such financing may also become erratic, lacking continuity.

Although it is difficult to sum up the main areas of research concentrated on by local administration, due to the diversity of performances, the main areas of financing are:

- Agricultural and biomedical sciences
- Technological research
- Economy and sociology

The Operation of Public Research Centres

According to the most recent OECD report (1992) on the S&T system in Italy, certain conclusions can be drawn regarding the characteristics of the operation of public research centres. According to this source, public research in these centres has a twofold aspect. On the one hand, there exist large centres, operating at a more than acceptable level and whose installations are up to the standard of the some of the best institutions in the world. On the other hand, there are an immense number of smaller public research centres which, in comparison with their larger counterparts, lack appropriate coordination of actions. The reason for this is that they usually depend on a Ministry with totally independent actions from that of the others. To palliate this question, the *MURST* initiated a series of actions to draw up sectoral plans, involving Ministries with common interests in such a way that the corresponding public centres could coordinate and plan their efforts along the same lines.

Another aspect pointed out by these reports is the type of research being carried out in these centres. In any scientific process, research may be classified in three basic types:

- Basic research
- Applied research
- Experimental development

In research centres, excessive importance is given to basic research over applied research. It is a known fact that basic research is one of the main guarantors of the continuity of an S&T system, but the orientation of research toward areas of greater interest (applied investigation) is of no lesser importance. According to the same report, Italy does not concede sufficient importance to the purpose based nature of research nor the vital need for results to convert into new products or innovation processes in industry. This aspect acquires special importance in the case of Italy. It must be kept in mind that the Italian economy depends greatly on its foreign sector, mainly dominated by important industrial companies which, in order to maintain competitiveness in the EU market, must be involved in a permanent process of renovation in R&D.

PUBLIC RESEARCH:
FINANCING AND MAIN PROGRAMMES

Before entering into an analysis of public expenditure in R&D and main programmes, a brief comment is called for on the position of Italy with respect to its neighbouring countries. The development of the economic and social background and its later integration with the other industrialized countries of the European Union have provided an important stimulus to scientific and technological research, and the need for public administration involvement. The Italian Government responded to the challenge, and provided society with the currently existing research centres, extending their lines of research in response to needs. In adapting to these needs, public research centres underwent significant reform.

Basically such reform has been oriented at rationalising the different areas of action in order to contribute to the modernisation of the country.

The programmes are directed at areas deemed to be of importance by the Government and which are of national interest. Such areas can be classified in five main groups, sometimes without a clear cut distinction between one and another:

- *CNR* determinant projects
- National programmes established by the *MURST*
- Space programmes
- International projects
- Sectoral plans developed by one or more Ministries

Another public activity is the provision of support to the production sector in certain research related areas, such as:

- Certificates
- Analysis
- Testing
- Quality control

Figure 4-6: Public R&D Funds 1987-1993 (%)

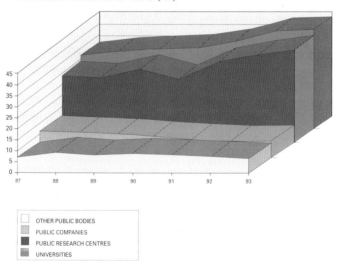

Source: ISTAT, 1995

Given that such tasks are not usually of great economic interest, are subject to strict control and demand a high level of impartiality, this means that often responsibility falls on the different levels of public administration.

The legal framework within which public involvement in topics related to scientific research is conducted, goes back to Act 283/1963 of 1963, which defines the responsibilities of the Interministerial Committee for Reconstruction (later to become the Committee for Economic Planning). This Committee granted the National Research Council (*CNR*) the power to carry out a report on the state of the Italian S&T system, in which the concept of scientific research expenditure was included for the first time. This Act has served as a reference to scientific activity in Italy up to 1989, the year in which the Ministry for University and Scientific and Technological Research (*MURST*) was created through Act 168/1989.

Figure 4-6 shows the destination of public R&D funds in the last decade.

The two most important facets of public research are the National Research Programmes and *CNR* Determinant Projects. Initially, the impression may be given that there is duplicity of responsibilities in the design of the programmes which cover the needs of Italian society, but this is not so. The differences between the two types of action are significant. While the majority of the former are aimed at the production sector (projects are usually developed in companies with the collaboration of public research centres), the latter are carried out in universities or public institutions.

National Research Programmes

A fundamental requirement of the S&T system is a solid government-laid basis in order to promote research; science in any country must be oriented toward socio-economic objectives of greatest interest. Although in the majority of industrialised countries most research projects are carried out in the area of production (around 60%), the role of the State consists in serving as a channel of funds and in addition to a normal percentage of investment (around 40%), redirect these projects toward areas of strategic interest.

As indicated in the previous section, the first National Research Programmes were passed in 1983 and started to be developed at the end of last decade, 1987 to be exact. High-priority lines were; microelectronics, biotechnology, health, construction, chemistry and metallurgy. These initial programmes, developed up to 1990, received 240 billion lira. It is important to note that the quantity approved by the Government to finance R&D activities was almost double the above figure, but appropriate use was not made of these resources. This situation was not new in the area of Italian science; there had been previous experience in the lack of resource allocation.

In 1989-1990, a new series of programmes was initiated which included; Bioelectronic Technology, Chemistry II, Advanced Materials and Environment. The necessary funding was 900 billion lira, the overall figure approved 1.7 billion.

The importance of the National Programmes in the funding of research activities is clearly reflected in the increasingly greater level of resources approved by the Government. Up to 1990, the total approved was over 1.9 billion. Of this figure, only a small fraction (328 billion lira) was actually allocated.

Such rapid growth in the quantities approved in comparison with the proportionally low quantities actually allocated, is an example of the inefficiency when locating funds in areas of greater interest. That is to say, in Italy, programmes are either not being implemented which are of sufficient interest to attract industry or public research centres, or are of scarce practical application.

Faced with this situation the Italian Government reoriented the areas identified within the National Research Programmes, which have evolved from support of traditional industries toward those of more technological content.

CNR Determinant Projects

The information available on *CNR* Determinant Projects is divided into three periods (1976-81, 1982-87, 1988-1993) corresponding to the first three generations of these projects. These are separated in groups according to the objectives. The first generation obtained funding of nearly 700 billion lira. The breakdown is shown in Figure 4-7.

Figure 4-7: Financial Distribution of the main *CNR* Determinant Projects

- Others 38%
- Energy 25%
- Environment 22%
- Industry 15%

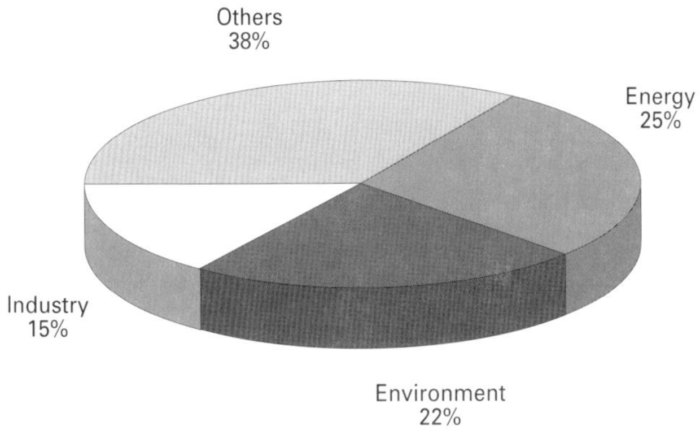

Source: CNR, 1994

During the eighties, the second generation of Determinant Projects came into effect and required a considerable increase in funding, which in the period 1982-1987 rose to a total of 1,000 billion lira. Almost half of this financing was dedicated to the programme of Advanced Technology, followed by Health and Energy. The environment project suffered an important cutback in its budgets.

**Figure 4-8: Second Phase of *CNR* Determinant Projects
Allocation of funds:**

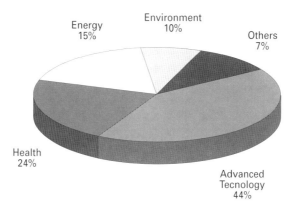

Source: CNR, 1994

The third generation of *CNR* projects, beginning in 1988, gave greater importance to the Advanced Technology and Energy Programmes, with a total of 66% of funds. The main problem was that Italy, like the rest of Europe, was immersed in a deep economic crisis, with a subsequent reduction of funds dedicated to R&D activities. In this period, total financing was only 350 billion lira.

In the period 1988-1993, the main part of the funds assigned concentrated on the Programme of Advanced Technology, with 40% of the total. This project embraced a wide span of actions, among which the following are noteworthy:

- Metallurgy
- Biotechnology
- Superconductors
- Optic-electronics
- Lasers
- Computer science.

The problem of these programmes is the same one as that pointed out previously for the National Research Programmes as a whole. To date, these have been unable to effectively attract the interest of the industry, and therefore the participation of the productive sector has been scarce in proportion to allocated funds. It can therefore be deduced that industry has not been able to take advantage of technological transfer to the extent initially expected.

As for the evolution of the funds, the budgets dedicated to the *CNR* Determinant Projects underwent important increases up to 1983, when the activities of the main national programmes began. From this point forward, cutbacks in *CNR* funds were successive, the reason being that the main part of capital is invested in the first phases of concrete programmes, and it was not until 1988 when investment was revitalised with the incorporation of the third generation

Table 4-1: Financing of Agricultural R&D by Public Administration and Semi-public Enterprise (1990-1991) - Millions of lira

Research organisations	1990	1991	Variation %	% of funding
Ministry for the University and S&T	291.789	308.383	6	39.6
Ministry of Natural Resources	123.207	126.836	3	16.3
CNR	66.293	105.528	59	13.6
National Agency for New Technology, Energy and Environment	40.347	38.192	-5	4.9
National Nutrition Institute	2.470	4.554	84	0.6
National Agricultural Institutes	12.293	18.392	50	2.4
National Biotechnology Institute	1.845	1.845	0	0.2
Agency for Seeds Selection	371	407	10	0.1
National Health Institute	1.523	3.784	148	0.5
Industrial Experimentation Station	11.183	11.187	0	1.4
National Institute for Fisheries Technology	5.306	5.371	1	0.7
Ministry of Merchant Navy	8.000	8.700	9	1.1
Regional	61.362	84.394	38	10.8
Subventions from International Organisations	11.180	11.380	2	1.5
Agency for the Development of the South of Italy	7.797	12.136	56	1.6
Total public sector	644.966	741.089	15	95.
Semi-public Enterprise	37.000	37.681	2	4.8
Total:	681.966	778.770	14	100.0

Source: MURST

of programmes. Another problem which occurred at the end of the 80's was the massive incorporation of personnel to the *CNR*, as well as the increase in the institution's running costs. All this forced a redistribution of the funds that the Government had provided to the *CNR*, consequently reducing Determinant Project funding.

After this brief overview of the main points in the history of *CNR* programmes, a description will be given of public projects in force until recently, and which in one way or another, have determined the evolution of the S&T system in Italy to a great extent. Such projects are grouped according to the sector in which they are operative.

Agriculture

Agricultural research in Italy involves a large number of institutions, both public and private, some of which develop research in the strict sense of the word and others which provide the technical and training support required by all scientific development processes.

The organisation with maximum responsibility for the planning of actions in this matter is the Committee for Economic Planning (*CIPE*). This Institution determines the conditions and financing of technological research, imposing the general lines to be followed and promoting the creation of specific programmes.

Coordination with the other National Programmes, fundamental in order to avoid overlapping in actions, is carried out by the Ministry of Universities and Scientific and Technological Research. However, at a technical level, coordination is the responsibility of numerous institutions, mainly at the ministerial level involved in the programmes. These Ministries and organisations are:

- Ministry for Natural Resources
- *CNR*
- Italian Agency for New Technologies, Energy and Environment
- Ministry of Industry
- Departments of Agriculture spread throughout the twenty regions of Italy

The main programmes developed related to agriculture are the following:

- The National Agricultural Plan, developed in the period 1987-1991
- Actions in the framework of the Special Fund for Applied Research and in that of the Programme for the Development of the South of Italy
- The National Forestry Plan
- The Three Year Environmental Programme
- The National Research Plan for Advanced Biotechnology
- Priorities outlined by the EU Framework Programme

The fundamental objective of research in these areas is the creation of new agricultural products in order to cover the growing market demand and thus achieve higher levels of competitiveness. High wage costs in Italy, compared with main competitors, have brought about the need for specific actions in order to reduce the maintenance and production costs of agricultural exploitation. Other research actions are; the study of non-food applications of agriculture; improvement in product quality; reduction in the deficit in the level of exports/imports (it should be borne in mind that the Italian agricultural market, and that of the EU in general, is faced with a situation of high competition); and the reduction of the impact of agricultural technology on the environment (study of plant diseases and creation of more resistant species).

The main action relating to agriculture was the *RAISA* (Advanced Research and Innovation in the Agricultural System) Plan of the *CNR*. This four year long Determinant Project was developed in the period 1991-1995. The main objective was to improve and conserve Italian agricultural resources, in line with market and environmental requirements.

The programme consisted of a series of sub-programmes, each of which was related to strategic research areas in the Italian agricultural system:

- Agricultural system and environment
- Agro-biotechnology in vegetable production
- Agro-biotechnology in animal production
- Biotechnology in food processing

The programmes were located in areas with special needs. Each problem area was assigned a series of research units coordinated by the Director of the Scientific Committee.

Total funds assigned to this programme were 226 billion lira. Personnel involved in activities was 400, the majority of whom were experts in different areas. The structure of programme execution basically corresponds with the structure of the Italian public S&T system: 60% of research was developed in universities; 20% in *CNR* institutes; and the remainder in enterprise through agreements.

The institutions, and financing which they contributed to the actions for the improvement of the agricultural system in Italy, are shown in Table 4-1.

ASTRONOMY

Among the countries being analysed, Italy has greater developed this area of research, a fact demonstrated by the degree of international prestige achieved in the international scientific community.

The main research activities in astronomy are presently being developed in the astronomical observatories, the National Research Council (*CNR*) and the framework of the universities. An idea of the scope of this type of project can be seen from the high level of financing received to date. In the case of the universities, this figure is around 12 billion lira, a quantity to which 5 billion more may be added for running costs.

Programmes developed within the *CNR* are carried out through two national groups (Astronomical and Cosmic Physics) which in turn are made up of seven public research centres. In addition to these, other centres also exist carrying out similar activities but which do not directly depend on the MURST. Examples of these are the National Astronomical Observatories and other institutions of the Italian Space Agency.

The main collaboration agreements and astronomical programmes to date are:

- Observatories of the US west coast.
- Observatory in the Chilean Andes (La Silla)
- Observatories of the Canary Islands (Roque de los Muchachos)
- Observatories in Hawai (Mauna Kea)
- Italian-American COLUMBUS project (Installation of a telescope in Arizona)
- *CNR* Radio-astronomy Programme
- International participation in ESA and other collaboration programmes agreed with the French counterpart of the *CNR*, the *CNRS*
- Participation in the construction of the LEST solar telescope

In total, personnel involved in activities is greater than 1,000 people, including scientists (about 600), technicians, administrative personnel, etc.

Biotechnology

Since the beginning of this decade, biotechnology programmes have been present in the national R&D plans as well as in the EU Framework Programmes.

In Italy, biotechnology funding is considerable. The development of specific programmes in this area is being simultaneously carried out by the *MURST*, the Ministry of Natural Resources, the *ENEA* and the *CNR*. Two *MURST* programmes presently in force are outlined below.

Table 4-2: *MURST* Biotechnology Programmes

Programme	Budget
Advanced Biotechnology	209 billion lira
Technology for Bioelectronics	99 billion lira
Total	308 billion lira

Source: MURST

The Ministry of Natural Resources coordinates the national programme, Development of Advanced Technology for Plants, with a total budget of 16.7 billion lira.

The *CNR* also develops three concrete initiatives in the field of biotechnology within the framework of Determinant Projects. (Table 4-3)

Table 4-3: *CNR* Determinant Projects - Biotechnology

Programme	Budget
Biotechnology and Instrumentation	84 billion lira
Genetic engineering	46.6 billion lira
Research and Innovation in the Agricultural System	226 billion lira

Source: CNR

Coordination of *CNR* activities is carried out through the National Committee for Biotechnology and Molecular Biology. Thanks to this organisation, set up in 1988, this series of Determinant Projects has reached a position of world wide prestige in this field. The reason for this is the wide experience of the *CNR* in developing projects of this nature. It should not be forgotten that the Biotechnology and Bioinstrumentation Programme was the first of its kind in Italy. The programme lasted five years, beginning in 1990 and concluding in 1994.

Table 4-4: Structure of bioindustry in Italy

New biotechnological companies	60
Companies already established	75
Total companies	135
Employees	2,400
ERD (millions $)	150
Returns on biotechnological products (millions $)	350

Source: Assobiotec, 1991

Table 4-5: Structure of Public Research in Biotechnology

Employees	
University	3,500
Other research centres	900
Total	4,400
ERD (millions of $)	
University and other Public Research Centres	190
Financing to Companies	50
Total	240

Source: Istat, MURST, 1989

Initial funding was 84 billion lira, employing over 250 researchers. The programme was eminently practical, proof of which is that the resulting applications and products are presently being used in industry.

Although the S&T system in the Italian productive sector will be analysed in greater depth further on, a short overview will be given here to the area of biotechnology, due to that fact that companies involved do not have the same high level of development as the national plans. Industry's interest in this area of research began approximately in the mid 80's and at the beginning of the 90's led to the publishing of two reports dealing with the development of Italian companies in this field. Of the two studies referred to, one was carried out by the Ministry of Scientific and Technological Investigation in collaboration with the National Committee of Biotechnology, and the other by a specially created committee by *Federchímica* (Italian Federation of Chemical Industries). A series of recommendations were contained in these publications:

- Substantial increase in R&D funding in the area of biotechnology.
- Improvement in the training of personnel
- Creation of new structures of technology transfer
- Development of new legislation to provide an appropriate framework for actions in this area

The reason for such recommendations can be easily explained by looking at the present structure of biotechnological research in enterprise. The type of company carrying out its work in this field is usually small (80% have less than 20 employees) and only a few actually carry out R&D activities. In fact, only 135 companies are involved in biotechnology, 40 of which exist with the help of the national R&D programs and grants from regional government.

The development of biotechnology in Italy is therefore almost exclusively the role of the State, both in research structures (universities, public research centres, the *CNR*, *ENEA*, etc.) as well as in the programmes carried out. The problem is not a question of low personnel levels in this area due to possible scarcity. Of the approximate 80,000 graduates per year from the 60 Italian universities, 30,000 have developed careers related to biotechnology in one way or another (medicine, engineering, pharmacy, biological sciences, etc.).

The differences between State involvement and that of industry in this field are reflected in Tables 4-4 and 4-5.

Environment

Environmental related projects are now a common aspect of the S&T system in industrialised countries. The European Commission includes a strong environmental component in most of its actions.

The Italian Government develops specific projects in order to research possible mechanisms which reduce environmental impact.

The motives for such actions can be summarised in the following points:

- Italy is one of the most densely populated countries in Europe. Its population is concentrated in the plains, coastal regions and around the large urban nuclei. In certain areas the ratio of inhabitant per square kilometre is 2,000.
- Italian industrial activity is one of the most important of the developed countries and the production of waste is estimated to be more than 40 million tons per year.
- Seismology is of special importance due to the peculiarities of the geologic aspects of Italy: in this century alone, the number of deaths due to earthquakes is 120,000 and economic damage, 119,000 billion lira.
- In the period between the 70's and the 90's the number of passengers (especially by road) increased by more than 100%.
- Agriculture and other primary sectors are of special importance in the Italian economy and the preservation of the environment is vital for the development of their activities.
- Efficiency in energy consumption in Italy is one of the lowest in the OECD countries.

This series of facts, in addition to others not indicated here, has urged the Italian Government to mobilise a high quantity of resources toward environmental areas. The development of these activities is deeply implanted in the national S&T system: the activity of Italian enterprise for example ranks fourth among EU Member States.

Of the two specific environmental programmes carried out to date, the first began at the end of 1989 and lasted for three years. This action, known as the Three Year Environmental Plan was approved by the Interministerial Committee for Economic Planning. The programme consisted of various subdivisions: General Programmes, Planning Guidelines and Strategic Projects.

The General Programmes dealt with: treatment of water; acoustic contamination; establishment of a national environmental information system; development of employment in areas related to environment, education and training.

Financial resources assigned to the programme were 9,668 billion lira, of which 4,415 billion was provided by the Environmental Ministry and the remainder by other administrations.

The second was the National Scientific Plan for Technological Research of the Environment, promoted simultaneously by the Ministry for the Environment and the Ministry for the University and Scientific and Technological Research. A unique aspect of this new programme was its intention, through a somewhat ambitious plan, to integrate the specific needs of an area with future needs and the technological innovation required in the mid-term.

Table 4-6: **Budget of the National Scientific Plan for Environmental Research and Technology**

Area of research	Funding (Billions of Lira)		
	Research	**Training**	**Total**
Reduction of traffic emissions	28.0	1.9	29.9
Reduction of emissions of energy production	38.5	3.2	41.7
Reduction of the impact of agriculture on the environment	29.5	3.4	32.9
Treatment of residuals	41.0	5.5	46.5
Supply of drinkable water	39.0	3.8	42.8
Non aggressive technology	8.0	1.0	9.0
Monitoring and administration	23.5	3.2	26.7
Total	207.5	22.0	229.5

Source: MURST

The programme started in 1991 through a first initiative called Programme for Environmental Training and Research. In trying to define lacking infrastructure and information, the plan identified the more important problems which had to be dealt with and subsequently established the main areas of action, which are as follows:

- Reduction of emissions resulting from traffic and energy production.
- Reduction of the environmental impact of agriculture and older industries.
- Treatment of residuals.
- Supply of drinking water.
- Development of non-aggressive technologies
- Environmental monitoring
- Environmental management

Opto-Electronic Technologies

The most important programmes in the field of opto-electronics began in the early 70's, when the Italian Government made the *CNR* responsible for the development of a series of projects within the framework of Targeted Programmes (equivalent to the actual Determinant Projects). The importance of these actions is two-fold; they were deemed to be of high priority by the organisations responsible for the public S&T system due to their strategic interest; since their initiation they have stimulated new channels of communication and technology transfer between public research and enterprise in Italy. The fact that industry's interest in the fields of opto-electronics was growing and that the State attempted to provide further impulse, has resulted in the creation of new synergy between the two scientific communities, presently being exploited in other fields.

Given that the nature of these projects is practical, countless relative industrial applications have been developed in a wide range of sectors; construction, transport, biology and medicine.

The first of the main opto-electronic programmes began in 1987 for a period of 5 years. The main objective was the development of new laser technology for both industrial and medical applications. This programme also brought about the above mentioned atmosphere of cooperation. In 1989, a new programme, foreseen for a period of six years was set up, in order to capitalise on this new linkage and create new areas of competition. When the programme finalised in 1994, the main objectives had been fulfilled:

- To create opto-electronic prototypes (and related components) with possible extensive areas.
- Make viable new and sophisticated technologies for Italian industry
- To continue stimulating these new channels of communication and cooperation between the public and private scientific community.
- To promote the training of new specialists in areas which in the future could develop new industrial applications or which could apply existing ones to Italian production.
- To disseminate best practices in opto-electronic within industry.

The opto-electronic programme was divided in two concrete actions (the first had a duration of two years and the second, three) and in 5 sub-projects. Funding was 52,952 billion lira. State-industry cooperation was obvious, given that 50% of funding was assumed by private companies.

The CNR Uffizi Project

Given the immensity of the Italian artistic heritage, various *CNR* research actions are aimed at the development of new techniques for the conservation of art. An example of such an action is the Uffizi Project which attempts to apply the knowledge gained in other branches of research to the treatment and study of works of art.

The programme was applied as a pilot experience in the Uffizi Gallery in Florence and consists of the following activities:

- Automated monitoring and control of environmental conditions in order to carry out continuous modifications and maintain climatic conditions appropriate to the conservation of art.
- Creation of a database containing information and documentation on the collection.
- Evaluation of the state of the collection; development of specific control and maintenance; creation of "clinical files" containing the conservation background.
- Telematic dissemination of the documentation stored in files, both for internal use by the museum as well as other research organisations (local, regional or foreign) and personnel.

The extension of the gallery is planned, taking advantage of the new technologies. For example, the computerisation of the filing system will provide new exhibition halls. The Uffizi Gallery was chosen as the pilot site given that it had already undergone restructuring and the incalculable value of its collection (not only in economic terms but also because it represents a great part of Italian and world history).

The Uffizi project, initiated in 1990, has been carried out through the following entities:

- Office for the Supervision of Artistic and Historical Heritage (Florence)
- Institute for the Research of Electromagnetic Waves. (Florence)
- Institute for Atmospheric Pollution (Rome)
- Electronic Engineering Department (University of Florence)
- SYREMONT Ltd, Milan.

Computer Science

Public research related to computer science has been developed almost exclusively under *CNR* coordination and financing since the beginning of the 90's.

Basically, actions have been within the programme Information Systems and Parallel Computation, carried out almost entirely in university centres and under *CNR* supervision.

In the period between 1989 and 1994, two phases of the project had already been executed. The first (1989-1991) was dedicated to the study of computer needs of the production sector and the definition of theoretical conditions for their resolution (theoretical and applied research). The second phase (1992-1994) focused on the creation of precompetitive prototypes that could be transferred to both science and industry.

The project was divided in 8 subprojects grouped in three areas in order to improve fundamental aspects such as the coordination of actions. (Table 4-7)

Table 4-7: Information Systems and Parallel Computation - Actions

Area 1:	Scientific Computation for Big Systems
Area 2:	Processors, Architectures and Languages
Area 3:	Software systems

Source: CNR.

Microelectronics

Microelectronics is acquiring vital importance in industry. The core of the electronics industry, it is also expected that microelectronics will become the primary industry world wide at the beginning of the next century.

Italy has been involved for some time in R&D topics related to this area, evident by the number of high technological companies located in the country which has resulted in close cooperation links between scientific research and the production of this type of product.

Examples of such companies, Italian and foreign multinationals developing activities in this field in Italy are:

- SGS THOMSON (Milan)
- TEXAS INSTRUMENTS (Rome)
- ERICSSON (Rome)
- ITALTEL (Milan)
- RADAR: ALENIA (Rome)
- OLIVETTI (Milan)

In addition to these, there are approximately another 100 companies involved in microelectronics related research, especially active in certain sectors of the market, having achieved important results. These small companies do not have sufficient resources in order to take on microelectronics projects on their own given that such projects require high spending so as to have extensive impact. On the other hand, the level of competition in the world market in this field is extremely high, the leaders being the United States and Japan.

For such reasons, the Italian Government, through the institutions responsible for scientific programmes, set up the following projects in microelectronics.

Funding for the Programme of Material and Devices for Solid State Electronics (MADESS) was 160 million dollars. The breakdown of initiatives and financing corresponded to; universities, 15%; industry, 25%; and *CNR*, 60%. The objectives of the project were to respond to the specific needs of industry, given that results had to be provided in a short period of time due to the rapid evolution of such areas of research.

The results of the programme were of great importance. The success of the actions carried out in industry resulted in the establishment of a new electronic community.

The positive outcome of the MADESS led to the initiation of a second phase with higher financing (250 million dollars), most of which was contributed by the Ministry for the University and Scientific and Technological Research. Its practical objective was to make use of the theories and prototypes resulting from the first initiative and apply them to the Italian production sector. The need to improve communication and cooperation links between the agents involved in this type of research was also stressed while special importance was attributed to SME participation to enable them to make use of the technology transfer and know how available.

The fundamental objective of the National Microelectronics Programme is to finance R&D programmes in Italian industry, the involvement of public research centres being only possible as subcontractors. The programme was set up in 1988 and to date two phases have already been developed. The first ran from 1988 to 1991, the second from 1991 to 1994 and the third, presently underway, from 1994 to 1999. The first initiative concentrated on binary and tertiary semiconductors, with an approximate financing of 140 million dollars. The second phase focused on bioelectronics with funding to the value of 120 million dollars. Financing for the third phase is 140 million dollars.

Table 4-8: National Microelectronics Programme (1994-1999). Main Areas of Research

Automotive
Telecommunications
Automation
Voice Pattern Recognition
Opto-Electronic
Biosensors

Source: MURST

Pharmacy

The Italian pharmaceutical industry is one of the most important in the world. According to 1992 data regarding employees, companies and profits, this Italian sector ranked in fifth position among the industrialised countries after the United

States, Japan, Germany, France and the United Kingdom. At present, more than 300 pharmaceutical companies are established in Italy, employing more than 70,000 workers.

It is not surprising therefore that this area is one of the most state of the art related to R&D. Of annual scientific production in pharmaceutical areas, more than 10% corresponds to Italy ahead of Germany (8.6%) and the United Kingdom (5.2%). The results obtained go beyond the mere economic power of these companies. The commercial effectiveness of such research is reflected in the resulting high level of profits: actual investment in R&D can be situated at 12% of total returns.

Although the rate of success is high, it should also be kept in mind that the investment level, either measured in global numbers or as a percentage of total spending, is proportionally lower than in other countries. Countries such as France invest more than 14% of returns and generally higher percentages of their GDP in R&D. It should also be considered that research, although providing good results, is not a key success factor of the Italian pharmaceutical sector. Comparison of the acquisition of patents by industrialised countries shows that Italy ranks in third place. The incorporation of new technology and products to be later marketed by Italian companies prevails over the development of R&D activities which could bring about similar results.

There have not been any national programmes in this area in recent years, although public administration has financed short term agreements between specific industries and public research centres or university departments.

A possible means of improving the performance of the Italian pharmaceutical sector in the world context would be to establish broad programmes and develop agreements between public administration and companies.

New Materials

Public research on new materials has been channeled through the Interministerial Committee for Economic Planning (*CIPE*). The *CNR* was entrusted with the development of a Determinant Project entitled Special Materials for Advanced Technology. The project was carried out in various public research centres and involved a total number of 200 researchers. An important aspect of the project was the first-time creation of communication links between the private and public sector. The *CNR* made available funding to the value of 85 billion liras.

The four subprojects of this project can be seen in Table 4-9.

In the first three years of operation, more than 200 operative units have been set up with a total cost of 85 billion lira.

Training was another project objective. Special networks for the exchange of knowledge and experience were created. In addition, special importance was given to the identification of new industrial activities whose main value-added was the use of sophisticated materials.

Table 4-9: Determinant Project of Special Materials for Advanced Technology - Subprojects

Advanced Ceramic Materials
Composite Materials
Materials with Special Magnetic, Electric and Electronic Properties
Material Characterization, Properties and Qualification

Source: CNR

Renewable Energy

The main public organisation in charge of R&D related to energy is the *ENEA* (Italian National Agency for New Technologies, Energy and Environment).

ENEA programmes are of various types. One series of projects is carried out entirely in the installations of the Agency, while other initiatives are carried out in collaboration with other public research centres.

The main areas of research are laid out in Table 4-10 although activities are focused on biomass and wind energy. Italian energy resources have also been mapped, identifying areas of preferable action.

Table 4-10: Main Areas of Research in Renewable Energy

Photovoltaic Energy
Wind Energy
Biomass Energy
Geothermal Energy
Hydrogen Cycle
Mini-Hydro Plants

Source: ENEA.

Telecommunications

Telecommunications have reached a peak in recent years. The proliferation of value added services is making possible new forms of business activity and greater efficient channels of communication. Optic fibre will be the new means of data transmission and therefore requires investments in R&D and infrastructure to ensure widespread implementation.

Italy has a network of more than 10,000 Km. of optic fibre with some sections still in the installation process.

Given these necessities, the *CNR* began a Determinant Project on Telecommunications oriented at the installation and development of plans and techniques for optic fibre cabling and wide band ISDN.

The development of telecommunication systems can never be carried out on an individual basis: the existence of numerous international standards calls for intense international collaboration. Research carried out in Italy follows the general guidelines laid down by the RACE Programme of the EU Framework Programme.

The projects are carried out fundamentally through the public research centres under the guidance of the *CNR* and also through specific collaboration agreements with companies involved in telecommunications. The financing of the Determinant Project on Telecommunications was over 68 billion liras. Project length is 5 years.

Table 4-11 details the main areas of activity of the this project.

Table 4-11: Determinant Project on Telecommunications

Subproject 1: Structure of Broadband Communication Networks

Network Architecture and Models
Equipment Functional Characteristics
Satellites for High Definition Television

Subproject 2: Technologies of Broadband Optical Communication

Coherent Optical Systems
Optical Switching
Low Cost Opto-electronic Component

Subproject 3: Terminals and Signal Processing in ATM Networks

ATM Multimedia Workstation
ISDN Multiservice Terminal
High Definition Television Studies

Subproject 4: Access Techniques for Broadband Networks

User Access Techniques
Broadband Techniques
Studies on Network Resource Management
Feasibility Study of an ATM Satellite System

Source: CNR

Project objectives are the following:

- Enable the Italian telecommunications industry to acquire the knowledge necessary to implement communication networks based on new technology in accordance with European standards.
- As the project is of a "horizontal" nature, (actions in very diverse disciplines of science), efficient coordinatation of the actions of R&D departments and public research centres is needed.
- Promotion of the training of new highly qualified technicians and specialists in this area.

Transport

In the last decade transport has acquired great importance in the normal development of society. Recent programmes have taken this into account. A wide range of organisations are involved in research tasks related to transport; universities, public research centres, the *CNR*, R&D departments of some companies, etc. It is important to point out that so far these projects have been developed on a more or less individual basis, through the initiative of each promoter. No single national entity took charge of the coordination of initiatives.

The first research programme dates back to 1982 with the creation of PFT1 (National Structures Transport Research). The main objective was to initiate actions of coordination at national level, although no concrete organisation was set up for this purpose. In the period 1982-1986, within the framework of PFT1, research contracts were signed with the *CNR*. The areas of action were later applied to the rest of the programmes. The following are the most noteworthy:

- The application of know-how in order to solve the main transport problems, making results available at national level.
- Financing dedicated to these activities responds to the established government policy, through the Interministerial Committee for Planning of the Economy.
- The coordination of human potential which initially belongs to different areas of science.

Given the success of the results, even without appropriate national coordination, the second phase of the programme (PFT2) was initiated in 1992 and ran through to the end of 1996. The 6 subprojects are outlined below:

- Subproject 1: Mobility Management and Planning Tools
- Subproject 2: Vehicles
- Subproject 3: Technological Systems and Infrastructures
- Subproject 4: Urban and Metropolitan Transport Systems
- Subproject 5: Freight Transport
- Subproject 6: Participation in International Transport Programmes

TECHNOLOGICAL INNOVATION IN ITALIAN INDUSTRY

Italian industry has undergone extensive restructuring since the beginning of the 70's. During that decade and the beginning of the 80's, smaller companies or SMEs experienced more dynamic growth, but eventually it was the big companies who were to occupy such a position until the current situation came about, in which the Italian market is dominated by large multinationals.

Analysis shows that the reasons behind this growth are characterised by the realisation of numerous innovative changes, which could give rise to the belief that they came about from intense research and substantial R&D spending. The reality is very different. In spite of the fact that innovative change has been intense, R&D expenditure is the EU average. The increase in production has been spontaneous throughout the last few years. The decentralisation of production and the introduction of changes in the productive system, aimed at specialisation, constitute the main innovations that have resulted in both competitiveness and production, without carrying out important spending on scientific research.

Truly competitive industry is impossible without scientific research. Italy took advantage of the transfer of technology from other countries, not in a passive way, but rather knowing how to adapt it to the needs of each sector and appropriately integrating it into usual operation. This has resulted in, for example, Italian industry being the second producer of mechanical elements in Europe. The productive sector, on the other hand, was able to adapt to the new demands of the internal market, especially those derived from emerging sectors of production (such as the textiles or wood industry), which are those sectors with higher degrees of specialisation.

This rapid adoption of foreign technology provided new channels for the training of more competent personnel, greatly increasing the technical human capital in the industrial sector.

This situation was circumstantial, although advantage was taken of it. Logically, greater innovative potential comes about as a direct result of scientific research, and only through this can sustained development be guaranteed in the near future.

R&D Expenditure in Italian Enterprise: Fundamental Indicators

The weight of Italian industry in national scientific research activity can be best appreciated by comparing percentage R&D expenditure in enterprise.

Table 4-12: ERD and % Distribution by Sector (1994)

	ERD (Millions of lira)	%
Public Administration	4,267,205	20
Universities	4,174,800	20
Enterprise	11,497,170	55
Non profit making institutions	901,565	5

Source: OECD, 1995

The table shows that the percentage of ERD carried out by enterprise is 55% compared with 40% carried out by Public Administration (spending by public institutions plus that of the universities). This structure demonstrates that the execution of the main part of scientific research corresponds to the production sector, initially responding to what is considered to be the correct spending struc-

Table 4-13: ERD carried out by Italian enterprise in the period 1987-1994

Year	Current lira (Millions)	Constant Dollars (1990) (Millions)
1987	6,689,889	5,738.7
1988	7,679,544	6,177.1
1989	8,698,468	6,589.1
1990	9,914,291	6,977.0
1991	11,039,721	7,216.6
1992	11,640,843	7,284.9
1993	11,751,523	7,044.1
1994	11,497,170	6,684.4

Source: OECD, 1995

Figure 4-9: Percentage growth in ERD in enterprise in the period 1987 - 1994. Constant prices

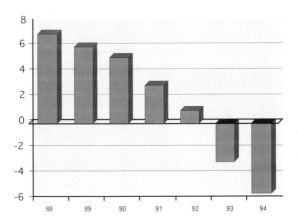

Source: OECD and ESIN, 1995

Table 4-14: Origin of R&D Funding in Italian enterprise
          Current Prices - Millions of Lira

	1991	1992	1993	1994
Enterprise	8,792,195	9,865,115	9,848,035	9,618,225
Government	1,299,162	1,148,917	11,98,205	1,139,908
Foreign	948,304	626,811	705,283	739,037
Total	11,039,721	11,640,843	11,751,523	11,497,170

Source: OECD, 1995

ture, given that it is the most common in the developed countries. As this sector is the main end user of research, the employment of most of the resources corresponds to it. The role of Public Administration must be that of a driving-force for scientific research through investment in strategic sectors, as well as coordinating and planning the lines that must be developed.

Although in current prices, spending by enterprise has progressively increased, the trend is contrary if measured in current prices. The situation is therefore unfavourable, as the R&D effort has continually decreased since 1993. The motive should be looked for in the Italian economic situation of that period. Europe was in the midst of a severe economic crisis and logically there were cutbacks in this type of spending. Perspectives are, however, good. The economic crisis is over, and the data points to a new increase in investment.

The percentage of growth in constant dollars is shown in Figure 4-9.

Data on the origin of the R&D funds in enterprise is shown in Table 4-14.

In constant prices, total investment has fallen which supposes a certain stability in current prices. As for the origin of funds, the evolution is similar. Except for funds coming from abroad which have experienced slight growth (due mainly to EU financing), the rest of the agents (public administration and enterprise) have reduced their contributions.

It can be deduced from the analysis of the breakdown of ERD according to activity (basic research, applied research, experimental development) that the public sector has not used the available funding to drive enterprise toward greater knowledge. Concentration has rather been on the maintenance of activities of greater relevance rather than new applications and improvement of existing technology. This gives rise to the feeling that the promotion of the most fundamental research has been entrusted to the public research centres and their collaboration with R&D laboratories in enterprise, rather than by their own initiative. An example of this type of collaboration is *CNR* Determinant Projects.

This situation is illustrated in Table 4-15. The data is somewhat outdated, which is understandable given the difficulty in distinguishing between the different types of research in addition to certain reticence by companies or institutions in providing such information. Nevertheless, it is still of use given that it is important to see the structure of fund applications, which is lasting in time, rather than concrete figures.

Table 4-15: ERD in enterprise according to the type of research carried out
Millions of Lira

	1987	1988	1989
Basic Research	429,203	474,913	524,321
Applied Research	1,407,164	1,568,789	1,767,809
Experimental Development	4,657,901	5,175,616	5,941,690
Total	6,494,268	7,219,318	8,233,820

Source: OECD, 1995

Table 4-17: Number of Researchers in Italian Enterprise (1990-1993)

	1990	1991	1992
Researchers	31,530	29,577	28,479
Technicians	23,285	22,355	21,920
Others	12,681	13,549	13,059
Total	67,469	65,481	63,458

Source: OECD, 1995

Table 4-18: ERD according to sector (Millions of lira)

Sector	1988	1990	1993
Agriculture	3,013	0	0
Mining	—	—	—
Industrial	6,925,631	8,874,665	9,911,615
Food	68,878	80,701	148,227
Textile	7,684	18,853	18,508
Wood, Paper	6,572	8,538	4,026
Petroleum, Chemical	1,749,934	2,207,659	2,454,228
Non Metallic Mineral Products	43,681	52,171	42,007
Metals	157,684	195,764	133,757
Metal Products	157,374	159,281	155,435
Machinery, Equipment	4,731,060	6,143,036	6,930,948
Other Manufacturing	2,764	8,662	24,479
Recycling	—	—	—
Electricity, Gas, Hydric Resources	193,004	288,397	517,651
Construction	20,636	10,935	5,145
Services	529,428	731,154	1,314,012
Sales, Commerce	—	—	2,480
Hotels, Restaurants.	—	—	—
Communications	41,955	16,699	224,002
Transport & Storage	3,096	600	0
Financial organisations	—	—	—
General Financial Activities	463,377	691,948	293,880
Communication, Society	21,000	21,907	23,350
Total	7671,712	9,905,151	11,748,423

Fuente : OECD, 1995.

Data relating to the number of researchers working in production does not differ substantially from the other indicators examined. It may be supposed that if total research is reduced, the number of researchers also falls. Table 4-16 reflects that the number of personnel dedicated to R&D activities progressively increased until 1990, and then diminished in 1991 and 1992. This implies that the process of technological innovation is less and less intensive in human capital, infrastructure and equipment. The same sources indicate that the number of researchers fell from 31,530 in 1990 to 28,479 in 1992, while the number of technicians was reduced from 23,285 to 21,920 in the same period.

Table 4-18 shows the relevant data for ERD according to sector.

The main sectors of the Italian economy carry out higher R&D investment. From Table 4-19, it can be seen that Food, Petroleum, Machinery and Services monopolize more than 80% of expenditure.

The recent fall off in investment has not affected all sectors in the same way. For example, in the services sector the level of resources assigned to R&D has increased while in the industrial sector it has diminished. This fact is reflected in Figure 4-10.

Figure 4-10: Evolution of ERD in Industry and Services
                (In millions of dollars and constant prices)

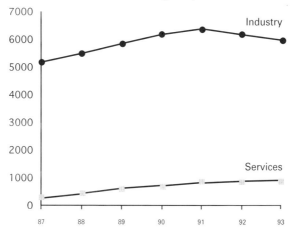

Source: OECD, 1995

The tendency toward a gradual increase in spending in the services sector is a common fact in the other lesser developed EU countries. However, this is not the case of R&D investment in industry. In Greece, Portugal and Spain spending can be seen to be growing in both cases, in a more and more rapid convergence process with the other countries of the environment, and not of divergence as in the case of Italy. It should be pointed out that investment levels are much higher in Italy than in the rest of these countries.

A fundamental aspect referred to at the beginning of this section was the transfer of foreign technology to Italian companies. The technology balance of payments (TBP) is used as an indicator of the "intangible" transfer of foreign technology (patents, inventions, designs, licenses, trademarks, knowledge, technical assistance, etc.). The TBP only covers part of the total technology transfer which includes the trade of equipment, the flow of information via the exchange of scientists and engineers, direct investment in other countries, etc.

Data available on the technology balance of payments indicates that throughout the last 30 years, Italy has shown an important deficit in technology trade with other countries, although in recent years this imbalance has been reduced with respect to the country's most direct competitors.

Overall data suggests the apparent maturity of the Italian production sector regarding the importing of technology, but there are certain deficiencies with respect to exports. Given that Italy's position in the use of knowledge is avant-garde, it is able to absorb the technology flow from abroad with the necessary speed and adaptation. Nevertheless, the export of know-how is not at the same level as that of its more powerful competitors, neither in quality nor quantity. There are various reasons for this, which can be summarised in the following two points:

- A lower research level than in other EU countries.
- A high degree of specialisation at the international level.

The consequence of this is that the Italian business sector has been able to create large multinationals represented in numerous countries with considerable levels of activity and with a high capacity to adopt new innovative processes, but with investment levels in R&D below the average of other large corporations.

Table 4-20 and Figure 4-11 shows the balance of payments in recent years.

Table 4-20: Balance of payments in Italian business
              (Millions of lira)

	1988	1989	1990	1991	1992	1993
Revenue	830,884	706,761	845,232	1,749,700	1,632,800	1,478,200
Payments	1,533,557	1,424,177	1,468,968	2,935,300	2,963,700	2,583,000

Source: OECD, 1995

Figure 4-11: Balance of Payments in Italian Business
             Payments/Revenue in the period 1984-1993

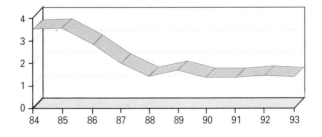

Source: OECD, 1995

The relationship between revenue and payments has stabilised in values around 1.7 of the average and it is not expected this value will change substantially in the short term. It can be deduced that numerous agreements of technological collaboration between Italian and foreign enterprise will continue to be signed.

The number of patents applied for usually serves as a clear indication of the intensity of technological change. From 1986 to 1992, the number of patents registered increased by 76%, one of the highest levels of growth in developed countries. In spite of this, in absolute numbers, the figure is still lower than that of other developed countries.

Table 4-21: Number of national patents - 1992

Japan	383,926
U.S.	185,857
Germany	98,940
France	78,753
United Kingdom	89,748
Italy	63,261

Source: OECD, 1995

Analysis of the evolution of the number of patents requested at the world level shows that the technological market has been significantly internationalised in the last 20 years. Italy has responded to this new challenge by increasing its number of patents abroad, although it is still far behind the larger industrialised countries. The gap is reduced if total expenditure by the Italian business sector is considered, which is not much smaller proportionally than the other countries. This fact is reflected in Figure 4-12, which shows the ratio of ERD to patents registered abroad in 1992.

Before moving on to a description of the technological innovation in this sector (companies carrying out R&D, size, average investment, resources dedicated, etc.) it should be pointed out that to identify technological change with R&D is a simplistic vision of the problem. Technological innovation is a much more complex phenomenon which implies diverse elements in addition to research and development, which in spite of their importance, are not the only source of technological change, though perhaps the most visible. Technological innovation is always measured with respect to R&D expenditure, although this relationship is neither lineal nor uniform in almost all sectors.

Figure 4-12: Business Patents /Expenditure in R&D.

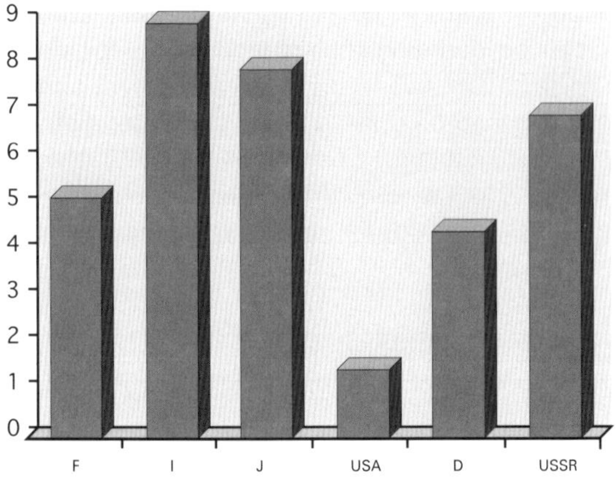

(Millions of Constant Dollars (1990) in the year 1992.

Source: OECD, 1995 and Own Elaboration

Numerous studies have been carried out in order to evaluate the process of technological innovation in Italian industry. Reference here is made to a study carried out in 1987 by the *CNR* in collaboration with the ISTAT, in which the research activity of a sample of more than 17,000 companies considered technologically developed was analysed.

Reference was made to the fall off in percentage of the average effort at each stage of the innovative process, smaller companies investing most of their resources in the acquisition of new infrastructure and less in R&D, while exactly the opposite occurred in big companies. The importance of R&D grows with the size of the company, while spending related to innovation, strictly speaking, is proportionally inverse. Small companies confront innovation via new plants and new machinery given that they do not possess the "technological capacity" to develop R&D processes by themselves.

Table 4-22: Cost of innovation developed by technologically developed companies in accordance with their size (%)

Number of employees	Type of cost		
	R&D	Engineering	Infrastructrure
20-49	7.4	15.1	73.1
50-99	9.6	17.5	68.2
100-199	10.6	18.3	66.2
200-499	14.3	16.1	64.2
500 -	21.4	29.5	43.5
Average	17.9	25.2	51.5

Source: CNR, ISTAT.

Table 4-23 shows the relative importance attributed to the decisive factors which lead a company to carry out changes of an innovative nature. On average, the acquisition of new machinery and infrastructure was the most important of all, closely followed by the design of new processes within the company. Other factors such as the proposals carried out by the departments of the company, clients' petitions, analysis of the competition, etc. were also considered.

Table 4-24 outlines total expenditure in the development of new innovative processes and the number of companies which have done so.

It can be seen that total R&D in SMEs is much lower than that carried out in large companies, even the total cost of innovation. This fact should be considered in the context that these SMEs constitute the main part of the business sector with percentages, of the total of companies, over 80%. This is also reflected in Table 4-25 in which the contribution of the SMEs to R&D and employment is compared.

Table 4-23: Decisive factors resulting in new innovative processes

Factor	Importance[*]
R&D (in-house or external)	2.1
Own engineering	3.1
Internal company proposals	2.3
Technology acquisition	0.5
Purchase of new raw materials	1.2
Purchase of intermediate products	0.8
Purchase of equipment	4.0
New specialized personnel	1.2
Training of personnel	2.2
Clients' demands	2.3
Collaboration with suppliers	1.6
Business synergy	0.3
Trade fairs, exhibitions	1.5
Collaborations with public research centres	0.3
Contracts with consultants and specialised companies	0.7
Analysis of the competition	2.0

[*] *Valued from 1 to 6*

Source: CNR/ISTAT

Table 4-24: Innovative companies by size and average cost of new product or product process

No. Employees	Total no. Companies	Average cost (Millions of lira)					Cost of Innovation (Billions of lira)
		-50	50-100	100-200	200-500	500-1000	1000-	
20-49	3,939	1,759	720	657	527	169	107	1,928
49-99	1,789	527	287	330	358	165	122	1,807
100-199	1,210	237	164	214	269	164	162	2,390
200-499	815	95	85	107	201	126	201	3,759
500-	467	28	24	41	77	63	234	19,406
Total	8,220	2,646	1,280	1,349	1,432	687	826	29,290
Percentage		32.2	15.66	16.4	17.4	8.4	10.0	100

Source: CNR, ISTAT

Table 4-25: Contribution of SMEs to R&D and Employment

Country	A	B	C
Belgium	33	36	69
Denmark	19	41	76
France	18	21	67
Germany	12	17	62
Netherlands	16	13	72
Ireland	82	—	83
Italy	19	23	81
Portugal	37	—	80
Spain	36	42	83
United Kingdom	9	6	65

A: SME expenditure on internal R&D of as a percentage of total enterprise expenditure on internal R&D.
B: Employment in R&D in SMEs as a percentage of the total R&D employment in enterprise.
C: Emploment in SMEs as a percentage of the total employment in enterprise.

Source: 2° Annual Report of the European Observatory of SMEs;
B & C: National data of the ENRS European Report on Science and Technology Indicators) 1994
1990 data (except: Belgium 1988; Germany and United Kingdom 1989; and Denmark 1991)

INTERNATIONAL PROGRAMMES: MULTILATERAL COOPERATION.

An important effort has been made in the last decade to reconcile Italian scientific research with that carried out by more developed countries, in the world context in general and in particular in the European Union. Such effort cannot be achieved only through the development of concrete plans inside Italy, but through the internationalisation of its scientific research, improving participation in new international forums as well as in the quality of research.

This necessity arises from the fact that S&T systems must face up to the new competition from countries which dedicate a larger quantity of resources to R&D as well as lesser developed ones, for which the only way to be technologically developed is through international cooperation programmes.

Italy's participation in the international framework of scientific research must be qualified as acceptable, considering the resources used in domestic R&D.

The legal framework within which collaboration is developed between Italian and foreign companies is Act 22/87 which addresses: association agreements for the exchange of knowledge; technological transfer; the realisation of research projects in collaboration; and the participation of public research centres in these combined programmes. The MURST dedicates a percentage of the funds for R&D promotion to sustain international projects wherever they may be carried out.

Main Multilateral Projects

The following table shows the most interesting initiatives of Italian research investigation, both due to the quality of the actions developed and the economic contribution.

Table 4-26: Main Agreements of International Cooperation

Participation in:
- CERN (European Organization for Nuclear Research)
- ERSF (European Synchrotron Radiation Facility)
- EMBL (European Molecular Biology Laboratory)
- ISIS (Spallation Neutron Source)
- ICGEB (International Centre for Genetic Engineering and Biotechnology)
- ESO (European Southern Observatory)

Italian Participation in CERN

Italy presently contributes 150 million Swiss Francs to the CERN, a centre specialised in the area of high energy physics, a figure which is approximately 16.7% of the institute's total budget. The main contributors are Germany (22.3%), France (16.9%) Italy and the United Kingdom.

Collaboration is carried out through the *INFN* (National Institute for Nuclear Physics) and through the National Committee, present in the CERN, established by the Ministry for University and Scientific and Technological Research. Of the total of full time personnel working in CERN (3,500) approximately 7% are Italian as are 28.5% of the 5,300 persons developing activities related to the centre.

Italian Participation in ESRF

In 1988, an agreement was signed by 11 European countries, among which was Italy, to set up the European Synchrotron Radiation Laboratory in Grenoble (ESRF). Italian participation is carried out through the *CNR*, the *INFN* and the Inter-universitary Consortium for Physics. Total financial contribution was 20 billion lira. A National Committee was set up to coordinate Italian actions.

Present Italian economic contribution to ESRF running costs (800 billion lira up to 1998) is approximately 15% of the total.

Italian Participation in EMBL, ISIS, ICGEB and ESO.

The EMBL is an international organisation consisting of 14 European countries and Israel. The headquarters is situated in Heidelberg with two other locations in Grenoble and Hamburg. Italian economic contribution to date has been 10 billion lira. The centre employs 250.

A bilateral agreement was signed with the United Kingdom for the use of ISIS due to the growing interest of the Italian scientific community in the study of the neutron spectroscopy and materials. The Italian contribution corresponds to 5% of the total.

Economic contribution to the European Southern Observatory is 25% of a total budget of 80 billion lira. ESO develops its activities through the telescopes located in the Chilean Andes. It is presently constructing one of the biggest telescopes in the world, with a diameter of 16 meters.

The fundamental objective of the International Centre for Genetic Engineering and Biotechnology (ICGEB) is the international promotion of research related to genetic engineering. It also carries out important support work in developing countries in aspects related to biology through training programmes in medicine, tropical agriculture, biotransformation of earth products, etc. The Centre has two main headquarters, one in Trieste and another in New Delhi. Almost all of the running costs of the centre are financed by the Ministry for Foreign Affairs and the Trieste Scientific Park.

EU Framework Programmes

The initial idea of the European Commission was to distribute financing for scientific research on the basis of country contribution to a common fund (principle of fair return), but time has shown that the assignments of projects has tended toward those work groups best able to carry them out, due both to their background as well as scientific knowledge. This implies that the objective of such financing is not that of cohesion between the S&T systems of the different countries and supposes a disadvantage for the lesser developed systems whose scientific base is not at the level of the developed countries.

The effect of this approach is partly compensated by another series of financing sources, also related to R&D, such as the Community Support Framework, ERDF, etc., which the European Union makes available to countries less favoured by the Framework Programmes. The imbalance is also compensated by the fact that the different countries are reluctant to establish cooperation agreements in areas of research which receive important domestic funding.

Analysis of Italian participation in the R&D Framework Programmes of the European Union, shows the following:

1. Regarding the industrial sector, where Italy has a predominant position in the European market, participation in the Framework Programme is balanced; revenue was approximately equal to contribution.

2. The technologies which could be called traditional with respect to scientific research have active participation (information technology, fusion, materials, transport, etc.)

3. Other areas such as new materials, telecommunications, non-nuclear energy, environment, electronic components and biotechnology have a lower participation by Italian projects.

The budgets in millions of ECUs (MECUs) and the III and IV Framework Programmes are shown in Table 4-27.

Italian participation in the III Framework Programme Figure 4-7) shows that financial contribution corresponded to 15% comparison with returns of just over 10% through agreements with companies, public research centres and universities. Data regarding the other countries is also given so that activity can be compared. It can be seen that the Italian contribution is substantially bigger, as are the returns obtained. However, the proportion between contribution and return is more favorable in other countries such as Portugal or Spain.

Projects have been carried out by scientific associations, public research centres, enterprise and universities. The distribution of approved projects among these institutions may be seen in Figure 4-14.

Table 4-27: Budgets of the EU R&D Framework Programmes III and IV.

	III		IV	
	MECUS	%	MECUS	%
1st Activity: Information & Communication Technology	2,516	38.1	3,045	27.68
Information Technology	1,532	23.2	1,932	15.71
Communication Technology	554	8.4	630	5.12
Telematic Systems	430	6.5	843	6.85
INDUSTRIAL & MATERIAL TECHNOLOGY.	1,007	15.3	1,995	16.22
Industrial & Material Technology	848	12.9	1,707	13.88
Measures and Rehearsals	159	2.4	288	2.34
ENVIRONMENT	587	8.9	1,080	8.78
Environment	469	7.1	852	6.93
Marine Science & Technology	118	1.8	228	1.85
LIFE SCIENCES & TECHNOLOGY	714	10.8	1,572	12.78
Biotechnology	186	2.8	552	4.49
Agriculture	377	5.7	684	5.56
Biomedicine and Health	151	2.3	336	2.73
ENERGY	1,063	16.1	2,256	18.34
Non renewable energy	267	4.0	1,002	8.15
Nuclear security	228	3.5	414	3.36
Thermonuclear fusion	568	8.6	840	6.83
TRANSPORT			240	1.95
Targeted Socioeconomic Research			138	1.12
TOTAL	5,887	89.2	10,686	86.88
2nd Activity: International Cooperation	126	1.9	540	4.39
3rd Activity: Dissemination & Evaluation	1% of the budget of each specific programme		330	2.68
4th Activity: Training & Mobility	587	8.9	744	6.05
TOTAL	6,600	100	12,300	100

Source: European Commission

With regard to the regional concentration of available resources of the Framework Programmes, it may be supposed in the case of Italy, given the gaps between the regions of the south (lesser developed) in comparison with those of the north (more developed), that funds would be concentrated in the latter. According to the statistics of the "European Report on Science and Technology Indicators" (1994), the regional concentration of these programmes is one of the lowest in Europe, behind only countries like Germany and Denmark. This can be explained by the fact that although in Italy research is concentrated in the North, in the other countries it is concentrated around a few urban nuclei, for example, the Ile de France or Rhone-Alps in France, or Madrid and Catalonia in Spain.

MAIN R&D INDICATORS

According to the last international comparisons published by the OECD, Italy was located among the first 8 countries of the world in 1991 regarding R&D expenditure (13,000 billion liras), a figure three times greater than that of Spain and sixty that of Portugal.

The four countries which dedicated higher funding to R&D were the U.S, Japan, Germany and France. Each one spent over 25 billion and the four together invested a total of 300 billion dollars.

Figure 4-13: Participation in III R&D Framework Programme % of participation in financing

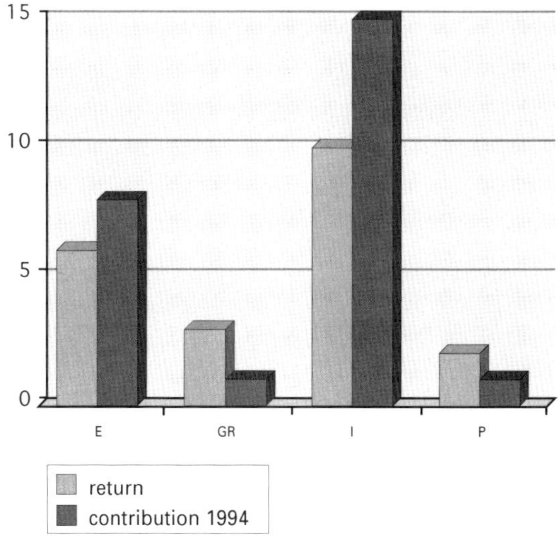

Source: European Commission, 1995

Three Asian countries (Korea, India and Taiwan) were ranked in twelfth and thirteenth positions with spending levels of over 2 billion dollars.

Figure 4-16 shows the evolution of the gross R&D expenditure in the period 1970-1991, Figure 4-17, the evolution of the percentage of GDP dedicated to R&D and Figure 4-18 the Italian position regarding the industrialized countries in the world.

Between 1980 and 1988 R&D expenditure experienced important growth (in constant lira) at an annual rate of 9.9%, a growth which logically was not uniform in the different sectors. At the beginning of the 70's, research promoted by

Figure 4-14: III Framework Programme. Italian participation.

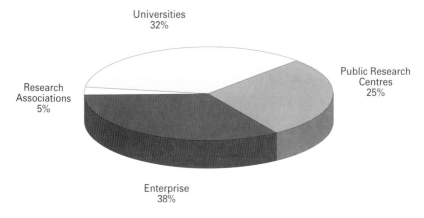

Source: European Commission.

Figure 4-15: Regional Distribution in Comparison with the Percentage of Industial Particpants regarding the National Total.

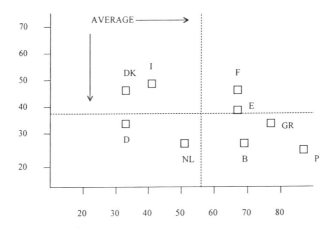

Source: The European Report in Science and Technology Indicators 1994.

the public sector was approximately of the same magnitude as that promoted by the private sector. In the last two decades, this situation has changed substantially and Italy has modified its participation structure in R&D initiatives in line with the most usual structures of developed EU countries (Table 4-28).

Table 4-29: Gross ERD according to Sector of Execution. Evolution from 1970 to 1991. (Billions of Lira)

Sector	1970	1975	1980	1985	1989	1990	1991
Public sector	252	517	1,187	3,932	6,102	7,087	8,690
Private sector	302	651	1,710	5,201	8,699	9,914	10,968
Total	554	1,168	2,897	9,133	14,801	17,001	19,658
Total in constant prices 1985	4,328	5,078	5,647	9,133	11,447	12,213	13,167

Source: Science and Technology Handbook, 1995

With respect to R&D financing, the results correspond to the application of funds. The main executioner of R&D is enterprise, but a higher percentage of financing corresponds to public administration.

Table 4-30: Origin and Application of R&D Funding in Italy 1994. (Billions of lira)

Origin of funds	Application of funds				
	Government	University	Private non profit making institutions	Companies	Total
Government	4,115	3,899	—	1,139	9,153
University	—	—	—		
Private non profit making institutions	—		—		
Companies	67	196	—	9,618	9,881
Foreign	84	78	—	739	901
Total	4,267	4,174	—	11,497	19,938

Source : OECD 1995

Distribution is shown in Figure 4-19 and Figure 4-20.

Figure 4-16: Evolution of Gross ERD in the period 1970-1991 (Billions of liras)

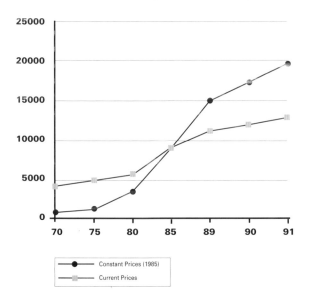

Figure 4-17: Evolution of ERD as a percentage of the GDP in Italy. Comparison with other EU countries.

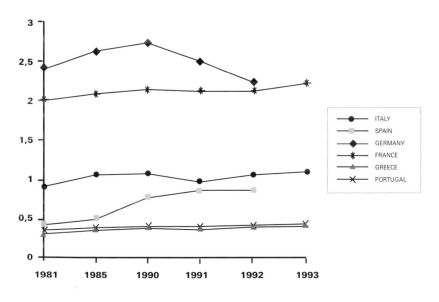

Source: OECD, EUROSTAT

The flows and composition of ERD are not identical. In the Figure 4-20, public financing is 42% (21% for activities in public research centres and 21% dedicated to universities). In companies, the R&D activities and innovation are practically self-financed. This demonstrates the scarce use of these funds by the private sector for scientific research. Only around 20% have their origin in domestic funds and 6% in foreign funds.

Table 4-31: Financing of Scientific Research.

Country	GDP (Millions of $)	ERD (% of GDP)	Industry Financed %	Government Financed %
USA	149,225.0	2.77	50.6	47.1
Japan	62,865.0	2.88	77.9	16.1
Germany	31,585.3	2.73	62.0	35.1
France	23,768.4	2.42	43.5	48.3
United Kingdom	20,178.3	2.22	49.5	35.8
Italy	*11,964.3*	*1.30*	*47.3*	*51.5*
Greece	336.3	0.45	19.4	68.9
Portugal	501.8	0.61	27.0	61.8
Spain	3,888.8	0.85	47.4	45.1
Belgium	2,751.5	1.69	70.4	27.6
Austria	1,796.7	1.40	53.2	44.3
Finland	1,541.8	1.87	62.2	35.5

Source: EAS (1991).

The financing structure shown in Table 4-31 is in line with the developed countries in the European Union, in which enterprise leads in R&D expenditure and financing

Another characteristic of the Italian S&T is related to the distribution of invested funds, according to the type of research carried out.

Table 4-32: Distribution of R&D Funds According to the Type of Project Carried Out (%)

Type of research	Italy	Japan	Belgium	Spain
Basic research	17.6	14.2	15.1	17.9
Applied research	43	27.6	30.2	36.8
Experimental development	39.4	58.2	54.7	45.3
Total	100.0	100.0	100.0	100.0

Source : OECD, 1995

Figure 4-18: Total ERD as a percentage of GDP, 1990 - Industrialised Countries

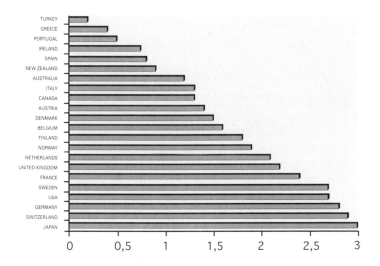

Source: The World Competitiveness Report, 1992

Figure 4-19: Execution of R&D Activities in Italy 1994 (%)

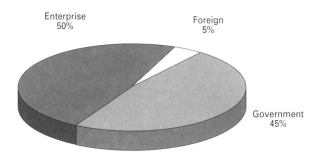

Source: OECD, 1995

Figure 4-20: Origin of Funding. Percentage Distribution 1994

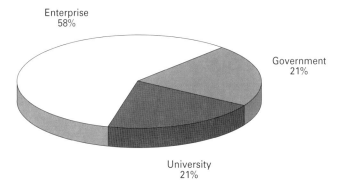

Source: OECD, 1995

The percentage dedicated to applied research is frankly superior to that of other countries with S&T systems similar to that of Italy. The reason is the higher concentration of resources in the evolution of the results of experiments toward more useful applications in national production. Recent trends show that this imbalance has not been corrected over time, but rather that differences have gradually increased.

The regional distribution of R&D funds also shows profound differences. In comparison with the high investment in the wealthy northern area, are the regions of the Mezziogiorno where capital invested in scientific research is much lower. Such areas as Basilicata, Emilia Romagna, Bolzano and Sardegna make up more than 76% of the total. This data can be observed in Table 4-33 which shows total expenditure in accordance with the socio-economic objective in the different Italian regions.

It is difficult to establish the number of personnel employed in R&D related tasks or innovation in general. Although figures exist in accordance with each source consulted, it is evident that there are big differences between researchers hired with a view to the realization of short term projects and those which take more time developing science related activities. The differences in supply and demand over the years have been significant and on most occasions circumstantial rather than as a result of the true capacity of the S&T system; on occasions figures have been greater influenced by the personal preferences of young people when choosing a career or that certain technologies evolve at a vertiginous speed, rather than by the adoption of concrete policies at the level of public administration.

From 1967 to 1992, researchers at full dedication increased from 61,291 to 142,855: in the last 25 years, the figure has doubled. This increase is more significant in the public sector than in the private, and in activities related to research rather than in support tasks (such as those carried out by technicians or auxiliary staff). In 1992, 57% of R&D personnel was hired by companies in comparison with 43% in the public sector (public administration and universities).

Table 4-33: ERD in Italian Regions in 1994. Billions of Liras

	Environ.	Infrast.	Construc	Health	Energy	Agricul.	Industry	Social	Non Oriented	Others	Total
Basilicata	400	1,940	2,819	0	0	1,840	800	0	550	50	8,399
Emilia Romagna	0	0	0	2,070	0	11,990	0	0	0	0	14,060
Lazio	0	0	1,270	0	0	0	0	0	0	0	1,270
Piamonte	0	0	0	0	0	0	150	100	400	0	650
Trento	123	0	452	0	0	0	0	40	150	0	765
Bolzano	400	0	0	0	0	12,760	0	800	600	100	14,660
Sardegna	0	0	1,000	0	0	16,573	0	0	0	0	17.573
Sicily	0	0	0	500	0	360	0	2,800	200	0	3,860
Toscana	0	0	0	0	0	0	0	5,324	0	0	5,324
Umbria	0	0	0	0	0	0	0	781	0	0	781
Fence d'Aosta	0	0	0	0	0	228	0	0	0	0	228
Veneto	0	7	850	0	0	3,151	0	68	0	0	4,076
Total	932	1,947	6,391	2,570	0	46,902	950	9,913	1,900	150	71,646

Source: Istat.

Figure 4.21: Evolution of R&D Personnel in Italy

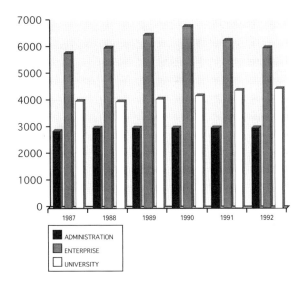

Source: OECD, 1995

Other differences exist between the research personnel employed in the public sector and those employed in enterprise. The first is in relation to the type of research carried out (basic research, applied research and experimental development). While in public administration, basic (especially in universities) and applied research are prevalent, the situation in enterprise is usually the opposite, the prevalence of experimental development accompanied by a high level of applied research, leaving aside the development of basic research. The second difference is to be found in the areas of science in which researchers are involved. In the public sector importance is given to human and social sciences, while this is not so in the enterprise sector. The disproportion in the number of researchers between both sectors is high.

CONCLUSION

Throughout this chapter, it has become clear that the Italian S&T system ranks among the more developed of the world. Although it is still not at the level of other countries such as the United States, Japan or the more developed EU countries, it is still in an advanced position. The system in general is compensated: the proportion of research carried out in the public and private sectors is practically the same as that of countries with state of the art systems.

Perhaps the main disadvantage with regard to competitors can be found in the type of activity carried out. The foreign sector is one of the most important of the Italian economy and is highly specialised in the trade of a series of very spe

Table 4-34: R&D Personnel Equivalent to Full-Dedication - Area of Science

	1987	1988	1989
Enterprise			
Human & Social Sciences	15	17	19
Experimental Sciences	57,463	61,632	64,925
Total	57,478	61,649	64,944
Public Administration			
Human & Social Sciences	3,646	4,057	4,110
Experimental Sciences	24,108	26,890	31,510
Total	27,754	30,947	31,510
University			
Human & Social Sciences	9,447	9,475	9,689
Experimental Sciences	33,496	33,594	34,353
Total	42,943	43,069	44,042

Source: OECD, 1995

cific products. A great number of innovative processes have been carried out, the necessary technology to do so being acquired from abroad leaving domestic research aside. This situation has resulted in the Italian economy being highly competitive in certain aspects, but at the risk of mortgaging future development to competitors.

Another negative aspect is the fall of R&D investment at the beginning of the 90's. After a period of rapid growth in the last 25 years, not only has expenditure stagnated in certain sectors, it has diminished in overall figures. It should be kept in mind, that the starting position of Italy was much more favourable than that of Spain, Greece or Portugal and that financing in countries such as France or Germany has been reduced.

Perspectives are good. The entrance of Italy into the European Single Currency (very probably in 1998 and in the company of Spain or Portugal) opens up new fields for competitiveness. The S&T system in Italy, a country with a long tradition in exporting, will be able to provide its production system with advantages in the face of other countries.

REFERENCES

OECD Economic Studies: Italy, 1996.

Informe general sobre la Actividad de la Unión Europea. Oficina de Publicaciones Oficiales de las Comunidades Europeas, 1995.

Europa en Cifras (EUROSTAT). Cuarta Edición.

Perspectivas de la Economía Mundial. Fondo Monetario Internacional, 1996.

Science and Technology Policy. Review and Outlook. OCDE, 1992.

Technology and the Economy: The Key Relationships. OCDE, 1992.

Statistiques de Base de la Science et de la Technologie. OCDE, 1995

Country Profile: Italy. 1995-1996. The Economist Intelligence Unit.

Reviews of National Science and Technology Policy. Italy. 1992. OCDE, 1993.

Scientific Research in Italy. Prof. Giorgio Sirili. Institute on Scientific Research and Documentation, CNR.

Le Attività e le Risorse per la Res Dell'operatore Pubblico in Italia, CNR. 1995.

Italian Science and Technology Handbook. Silvio Dottorini. Embassy of Italy. Canberra.

Scientific Research in Italy. Prof. Giorgio Sirilli. Euromecum, August 1996

Commision of the European Communities, Science and Technology, Public Understandig and Attitudes, June 1993, Brussels

R&D Evaluation in Italy: A Science and Technology Policy View, Research Evaluation, Volume 5, Number 1, Silvani A., Sirilli G.,April 1995

Scienza e Tecnologia in Cifre, 1993. Isrds-Cnr

La Promozione delle Scienze Trasferibili: Problemi Organizzativi e risultati nel caso delle biotecnologie, Rapporto tecnico n° 11/1994. Isrds-Cnr, Roma

La Spesa nell'industria per la Ricerca Scientífica. Centro de Studi Cofindustria. Sipi, Roma

I Programmi Nazionali di Ricerca, Rapporto Tecnico n° 13/1994 Isrds- Cnr Roma

L' impegno dell'industria nella ricerca. De Marchi M. Napolitano G. Rocchi M. Rapporto Tecnico n° 2/1993 Isrds- Cnr, Roma

The Developement of Transfer Science. Stanford, 1994

Networks of innovators: a Synthesis of Research Issues. Research Policies. Freeman C.

The National System of Innovation: Italy. Malerba F. Oxford University Press 1993.

Economic and Social Impact of National Research Programmes. Presented at the conference on Technological and Economic Development. Le Mans 14 de October de 1994

La scienza in Italia negli ultimi quaranta anni. Franco Angeli Editore, April 1992 .

Università e Ricerca nel e per il Mezzogiorno Franco Angeli Editori, March 1989.

APPENDICES

APPENDIX I

GENERAL INFORMATION

SPAIN

Basic Data

Area:	504,750 Km², of which 30% is arable, 12% is pasture and 31%, wooded areas.
Population:	39 million (1991 census)
Main Cities:	Madrid: 3,010,000
	Barcelona: 1,643,000
	Valencia: 753,000
	Seville: 683,000
Climate:	Mediterranean in the east and south; temperate in the north and west.
Currency:	Peseta. Average exchange rates in 1994: 134 ptas=1 $; 82.6 ptas=1 Deutschmark M; 158.5 ptas=1ECU

Economic Structure

The Spanish economic structure underwent a radical change in the period after the Second World War and in particular after the stabilisation plan of 1959 when the Spanish economy began to open up. From this point onward, the foundations for industrial reconstruction were laid down, while the continually growing and emerging tourist industry resulted in the creation of a strong services sector. An economic structure based mainly on agriculture and fishing was gradually abandoned.

Table 1: Main Economic Indicators
           (1994)

Real growth in GDP (%)	2.0
Unemployment rate (%)	24.2
Inflation (%)	4.7
Public Deficit (% of the GDP)	6.7%
Current account balance (% of the GDP)	-1.1%

Source: Economic Intelligence Unit, 1996

Table 2: Comparative Economic Indicators
           (1994)

	Spain	France	Germany	Italy	UK
GDP ($ bn)	485.2	1,335	2,050	1,020	1,019
GDP per capita ($)	12,380	23,050	25,280	17,900	17,480
Inflation (%)	4.7	1.7	2.9	3.9	2.6
Current account balance ($ bn)	-4,0	9.0	-25.0	15.5	0.3
Exports ($ bn)	74.1	216.0	380.5	189.5	206.8
Imports ($ bn)	88.6	208.0	340.0	154.1	232.2

bn = billions

Source: EIU, 1996

The contribution of Spanish industry to the national GDP reached 24.4% in 1994. In the 60's and at the beginning of the 70's, when Spain was under the Franco regime, a protectionist policy was implemented in industry, thus isolating the sector from external competitors. With the arrival of democracy in the 70's and given the world wide economic crisis, profound restructuring of the sector was called for. The process which was abruptly accelerated on Spain's entry into the European Community in 1986 when the country was immersed in a new market where competition was difficult. A large number of firms were taken over by large multinationals, attracted by the low wage costs and opening up of markets. Spain was an excellent platform from which to establish industries of all types and to market products inside the Common Market. This "new re-industrialisation" as a result of foreign investment enabled the Spanish industrial sector to rapidly converge with the other European Member States, although much still remains to be done. An appropriate example is that of the motor industry in which Spain ranks third in exports in Europe (behind France and Germany) and fifth in the world when in fact no national manufacturer exists.

The services sector accounted for 63.4% of GDP, indicating its tremendous weight in the Spanish economy as a whole. Within the sector, the tourist and financial sectors (especially the banks) are of particular importance. Revenues from tourism contributed in 4.5% to the GDP, approximately 50% more than the national trade deficit. Spanish banking is not particularly of international importance as far as the volume of transactions is concerned, but profit levels are among the highest in the world and the growth rate is high. With regard to the future, it is expected that services related with telecommunications will acquire special importance.

Agriculture and fisheries are being relegated to a second level status in an economy increasingly more oriented toward industry and services. In 1994 they accounted for only 3.7% of GDP. Nevertheless, Spain still remains the European leader as regards fishing (it should also be kept in mind that coastline is proportionally large with respect to the country) while in the case of agriculture, large horticultural holdings supplying markets in northern Europe are to be found in the southern regions due to their excellent climate.

Regions

The phenomenon of regions is not new in Europe, but in the case of Spain it takes on greater importance, due not only to the different cultural and administrative facets of each one, but also because of economic and structural divergences. While the GDP per capita of certain regions such as Madrid, Catalonia, Navarre and the Balearic Islands is higher than the EU average, other regions such as Andalusia or Extremadura do not reach 55%. In general, richer regions (with the exception of Madrid) are on the Mediterranean coast and in the area of the basin of the river Ebro, while the south, centre and northwest are most depressed. The industrialised regions are Madrid, Catalonia and the Basque Country, the last having gained importance in recent years as a result of high industrial reconversion through the modernisation of its obsolete industry.

Foreign Relations

Spain's international relations are dominated by its membership of the EU which the country joined in 1986 and which has been one of the key factors for recent national development. Political consensus on European integration is total, despite some current complaints as a result of restructuring in agriculture and fisheries.

The country's historical bonds with Latin America and the north of Africa are of special importance. Spain has endeavoured that such links result in foreign trade and cooperation agreements, while the country also serves as a valuable interlocutor for the EU with these countries, whose markets will be of vital importance in the future, in spite of the reigning political instability.

PORTUGAL

Basic Data

Area:	92,082 Km² of which 34% is arable, 9% is pasture and 36% forestry.
Population:	9.91 million (1995 census)
Main Cities:	Lisbon: 2,062,300 Oporto: 1,262,200
Climate:	Mediterranean in the south and temperate in the north.
Currency:	Escudo. Average exchange in 1995: 110 Escudo = 1 $. 104 Escudo = 1 Deutschmark

Economic Structure

The economy in Portugal has experienced practically the same evolution as that of the other countries of similar background: in the last 25 years its economy has progressed from being agriculture and fishery based to an economy dominated by the increasingly more powerful sectors of services and industry. It should perhaps be noted that the competitiveness of industry is based on low wage costs. Portugal's entry into the European Community favoured the growth of both sectors, industry being dominated by textile manufacturing, and services by tourism and retail trade.

The lack of incorporation of new technology into agriculture and fisheries resulted in a fall off in competitiveness with respect to Spain and France. In addition, Portugal was seriously prejudiced by restrictions in fishing catches to the benefit of surrounding EU countries. As in the case of Spain, Portugal has an immense fishing fleet.

Traditionally the main Portuguese industry is textiles, which experienced a state of crisis as a result of the level of competition established by Asian countries. Nevertheless, the incorporation of new organisational procedures and the introduction of new technologies has enabled this sector (vital for the domestic economy) not only to survive but to recuperate. Although not particularly competitive in any one area of industry, Portugal has been able to create niches of industrial activity where it is outstanding in relation to other countries; ship repair, paper production, production of moulds, etc.

Recently a growing number of foreign multinationals have set up delegations in Portugal, taking advantage of the low salary costs and using the country as a launching pad for the commercialisation of their products in Europe. Government backed support for the establishment of industries of high technological level is also on the increase.

Economic policy of the Portuguese government is primarily guided toward the fulfilment of the criteria established by the Maastricht Treaty, thanks to which a further step will be taken toward European integration. Much has been achieved and it is quite probable that the conditions will be reached in their entirety, given that the Treaty is a platform for high economic growth in the future.

Table 3: Comparison of Main Economic Indicators (1994).

	Portugal	Spain	France	Italy	Germany
GDP ($ bn)	77	485.2	1,355	1,020	2,050
GDP per capita ($)	7,800	12,380	23,050	17,900	25,280
Inflation	5.2	4.7	1.7	3.9	2.9
Current account balance ($ bn)	-1,0	-4.0	9.0	15.5	-25,0
Exports ($ bn)	18,5	74.1	216.0	189.5	380.5
Imports ($ bn)	25,2	88.6	208.0	154.1	340.0

bn = billions

Source: EIU. 1996

Regions

Unlike neighbouring Spain, Portugal is not a quasi-federal state where regions can implement their own economic and development policies. Such policies are carried out under the general guidelines laid down by the Government in Lisbon. Such centralism avoids the type of problem suffered by Spain in achieving adequate financing for its Autonomous Regions. A characteristic difference among Portuguese regions that should be highlighted is the level of prosperity between the coastal regions and those situated in the interior. While the large coastal cities monopolise most of domestic economic resources, the other regions have a much lower development. An example is the area of Antalejo where GDP per capita does not even reach 20% of the EU average. Such gaps, instead of being reduced have widened. According to Government studies, in the year 2,000 more than 70% of the GDP will be contributed by coastal regions and only 30% by interior regions.

Foreign Relations

Portugal's link with its Spanish neighbour is an important factor in its foreign relations. Such links are cultural as well of common interest, especially when establishing a common front within the European Union in order to avoid the pressures of more powerful countries within this common space. Portugal has known how to take advantage of the weight of Spain in order to obtain greater financing from the European Commission, especially in the case of the Cohesion Funds, in which Spain, Greece and Ireland also participate. However, there are also numerous areas of contention in topics of importance to both economies (fishing or agricultural quotas) which separate the two countries.

Another fundamental aspect is the exchange and trade agreements with its former colonies. Special mention should be made of important existing agreements, for example with Brazil and with former colonies in eastern Asian.

APPENDIX 1

GREECE

Basic Data

Area:	131,940 km² of which 23% is arable: 8%, crops; 40%, pastures and 20% forest.	
Population:	11 million (1993 census)	
Main Cities:	Area of Athens:	3,095,775
	Thessaloniki:	739,998
	Patras:	172,763
	Heraklion:	127,600
	Larissa:	113,426
	Volos:	106,142
Climate:	Mediterranean	
Currency:	Drachma. Average exchange in 1994: 273 drachma =1$; 158 drachma =1 Deutschmark; 309 drachma =1 ECU	

Economic Structure

The Greek economic model is capitalism combined with a public sector which contributes 60% of GDP. According to World Bank estimates, the gross national product (GNP) in 1992 was 75,106 million dollars (average prices 1990-92) and income per capita was 7,180 dollars. Greece is member of the European Union as well as the Organization for Economic Cooperation and Development (OECD).

The livestock-agriculture sector ensures that the country is practically self-sufficient, apart from meat and diary products. In 1992, agriculture (including farming, forestry and fishing) attributed 14.9% of the GDP, providing employ-

Table 4: Main Economic Indicators (1994)

Real growth in GDP	1.2
Unemployment rate	9.6
Inflation	15
Public Deficit (% of GDP)	9
Current account balance (% of the GDP)	-2

Source: EIU, 1996

Table 5: Comparativve Economic Indicators (1994)

	Greece	Spain	Germany	Italy	UK
GDP ($ bn)	82,9	485.2	2,050	1,020	1,019
GDP per capita ($)	8,200	12,380	25,280	17,900	17,480
Inflation (%)	15	4.7	2.9	3.9	2.6
Current account balance ($ bn)	-14	-4.0	-25.0	15.5	0.3
Exports ($ bn)	6.8	74.1	380.5	189.5	206.8
Imports ($ bn)	21.5	88.6	340.0	154.1	232.2

bn = billions

Source: The Economist, 1996

ment to 21.1% of the population. During the period 1980-91 this contribution to GDP increased at an annual average rate of 0.2%. Industry (including mining, manufacturing, energy and construction) attributed 26.1% of the GDP in 1991 providing employment to 27.5% of the labour force. In the period 1980-1994, the industrial GDP increased at an annual accumulative rate of 1.2%. The remainder of the GDP is attributed to services.

Tourism has been an important source of revenue for Greece during the last three decades. The sunny climate, abundance of beaches, history and natural beauty combined with low prices and the continuous improvement of transport and accommodation make the country a favourite holiday destination. In 1968, Greece was visited by one million tourists, a figure which in 1991 had ascended to 8.03 million, with resulting profits to the value of 2,566 million dollars.

In spite of austere measures taken in the 80's and at the beginning of the 90's, one of the biggest structural problems of the Greek economy is the public debt. Great part of this situation has been attributed to administrative inefficiency in the public sector which still controls approximately 60% of overall economic activity, absorbing more than half of the revenue obtained via taxes, and at the same time sustaining an expensive social welfare system. A substantial improvement in economic indicators was expected from the privatisation programme implemented by the Government at the beginning of the nineties, but as a result of political discord, prospective benefits were quite lower.

Another problem is that of the existence of a second parallel economy: an immense submerged economy that obviously cannot be quantified and which constitutes a permanent source of fiscal fraud, preventing greater national development. Various attempts have been made by Public Administration to put an end to this, but given the immense political cost that this would suppose, attempts have been fruitless.

Based on low salary costs and tax benefits, Greece has tried to attract foreign investment in an attempt that this serve as a new motor to the economy. The success of this initiative has been important but not to the extent expected given the bureaucratic obstacles faced by the companies, in spite of the simplification of operating procedures (new legislation, incentives, etc.).

Greece received 4,060 million dollars (equivalent to 5% of GDP) in 1992 through the EU Structural and Cohesion Funds. In 1993 the European Council approved a five-year plan, (the Delors plan) to facilitate the convergence of the Greek economy with those of the other EU member states. This plan foresaw the gradual reduction of annual inflation to a level of 4% (at present it is 9%) and public debt.

Regions

Regional differences exist of various types. On average, Greek population density is low, although not so much in the regions of the north, in particular that of Attika where it is quite high, while the level is inferior in the south and islands. This data reflects the strong trend of population migration from the south to the large urban centres located in the wealthy north. Average income per capita is only half the European Average, Athens being the richest area, while the islands are the poorest regions. Employment in the agricultural sector (40% of the total) is three times higher than that of Europe. Industry is concentrated in the two large urban areas of Thessaloniki and Athens, while services, specially tourism, is located in Athens as well as in the Aegean islands (50%). Regional economic activity is clearly lower than the European average, with the exception of Athens.

ITALY

Basic Data

Area:	294,060 km² of which 41% is arable, 17% pastures and 23% wooded.	
Population:	57.2 million	
Main cities:	Milan:	1,371,000
	Rome:	2,693,000
	Naples:	1,055,000
	Turin:	962,000
Climate:	Mediterranean.	
Currency:	Lira. Average exchange in 1994. 1.612 lira=1 $. 1.909 lira=1 ECU. 995 995 lira = 1 Deutschmark	

Economic Structure

Italy is an importer of raw materials; both energy and agricultural products come from abroad. In the case of services, the country is not really competitive at world level with the exception of tourism. The main source of wealth is to be found in industry, one of the most important not only in Europe but in the world. The number of large multinationals is not high but companies like Fiat, Olivetti or Fininvest are of importance in the European economy. The main contribution to the Italian economy is made by small companies producing consumer goods. Of special interest are the white good and textile industries, the latter producing high quality design and products. A typical success story of this type of company is Luxottica, a world wide leader in the sale of glassframes.

Table 6: Main Economic Indicators (1994)

Real growth in GDP	2.2
Unemployment rate	11.3
Inflation	4.0
Exports ($ bn)	189.5
Imports ($ bn)	167.5
Public deficit (% of the GDP)	9.0

bn: billions

Source: EIU. 1996

Government participation in the economy is extremely important. A high number of companies have public participation. However, as a result of cutbacks in state aid to firms, favoured by the EU restrictions, the role of the state in these companies is being reconsidered.

Economic policy in the last 20 years has been characterised by the contrast between responsible monetary policy, under the direction of the Bank of Italy, and a budgetary and fiscal policy conditioned by political instability, a permanent characteristic of the country. The reality is that Italy is basically sustained by a highly professional public administration and not so much by the politicians in charge of running the country, who are not usually long in office.

As already indicated, the industrial sector is the main support of the Italian economy. Approximately three fourths of national exports correspond to this sector. Services, responsible for another substantial fraction of Italian exports, has the added advantage of contributing greater added value than manufacturing, although its activity is usually related to the supply of services to industry.

The high level of exports of Italian industry has resulted in heavy dependence on the balance of trade. This has been characterised by fluctuations in their final values throughout the years, strongly influenced by the exchange rates of the lira and economic cycles.

Although Italy possesses a high market quota at world level in consumer goods, it is not specially competitive in state of the art technological products. Their competitive advantage rests especially on new innovative processes rather than on purely scientific research. There are some exceptions such as the large multinationals already mentioned, the telecommunications sector or the aerospace industry. The textile sector, where technology application is up to the standard of that of any other country, should also be mentioned, explaining the special competitiveness of the country in this traditional sector.

As for the primary sector, Italy shows a high deficit in agricultural products and livestock. Although there is still wide margin for the improvement of this type of exploitations, the reality is that the national territory does not have sufficient capacity to supply a population of nearly 57 million inhabitants.

Table 7: Comparative Economic Indicators
(1994)

	Italy	France	Germany	Spain	UK
GDP ($ bn)	1,020	1,335	2,050	485.2	1,019
GDP per capita ($)	17,900	23,050	25,280	12,380	17,480
Inflation	3.9	1.7	2.9	4.7	2.6
Current account balance ($ bn)	15.5	9.0	-25.0	-4.0	0.3
Exports ($ bn)	189.5	216.0	308.5	74.1	206.8
Imports ($ bn)	154.1	208.0	340.0	88.6	232.2

bn: billions

Source: EIU. 1996

Regions-The Problem of the Mezzogiorno

While Italy endeavours to achieve a better position in the European Union, the regions of the south known as the Mezzogiorno (Abruzzo, Molise, Campania, Apulia, Calabria and Basilicata in addition to the islands of Sicily and Sardinia) are the poorest in the EU Member States. They make up approximately 40% of the national territory and 36% of population. The rate of unemployment is 25%, double the figure in the north or in the central regions. Contrary to the rest of the country, the most important sector here is agriculture, ahead of an industrial sector which has been totally stagnant for more than a decade. In this same period, the only sector which has evolved positively has been services. This series of circumstances was further strengthened by policies which limited themselves to patching up a situation which demanded another type of treatment.

APPENDIX II

RELEVANT ADDRESSES

SPAIN

COMISIÓN INTERMINISTERIAL DE CIENCIA Y TECNOLOGÍA.
DIRECCIÓN GENERAL DE INVESTIGACIÓN Y DESARROLLO
C/ ROSARIO PINO 14/16
28020 MADRID
TEL: +34-1-3360400   FAX:+34-1-3360435
E-MAIL: sgnid@cicyt.es

CONSEJO SUPERIOR DE INVESTIGACIONES CIENTÍFICAS (CSIC)
ADMINISTRACIÓN CENTRAL
C/SERRANO, 117
28006 MADRID
TEL: +34-1 -585 50 01   FAX: +34-1- 411 30 77

INSTITUTO DE ASTROFISICA DE CANARIAS (IAC)
C/ VÍA LÁCTEA S/N
E38200 - LA LAGUNA
TENERIFE - ISLAS CANARIAS
TEL: +34-22 605200   FAX:+34 -22 605210   TELEX: 92640
E-MAIL: postmaster@iac.es

CIEMAT (CENTRO DE INVESTIGACIONES
ENERGÉTICAS, MEDIOAMBIENTALES Y TECNOLÓGICAS)
AVDA COMPLUTENSE Nº 22
28040 MADRID
TEL:+34-1-3 46 60 00 / 01   FAX: +34-1-3 46 60 05.

MINISTERIO DE INDUSTRIA Y ENERGIA
GARCÍA DE PAREDES, 65
28071-MADRID.
TEL:+34-1-442-86-33/537-17-81.   FAX: 34-1-399-25-33.

DIRECCIÓN GENERAL DE INVESTIGACIÓN
Y DESARROLLO (MINISTERIO DE EDUCACIÓN Y CIENCIA)
CALLE ROSARIO PINO, 14 - 28071-MADRID
TEL: +34-1336-04-00   FAX:+34-1-336-04-35

Telephones numbers in Spain are due to change in 1998

INSTITUTO NACIONAL DE CALIDAD Y EVALUACIÓN (INCE)
CALLE SAN FERNANDO DEL JARAMA, 14
E28002 MADRID
TEL: +34-1-562 54 00   FAX: +34-1-561 89 21
E MAIL: info@ince.mec.es

UNIVERSIDAD COMPLUTENSE DE MADRID
SERVICIO DE INFORMACIÓN
AVDA. DE SÉNECA Nº 2
28040 MADRID
E MAIL: infocom@rect.ucm.es

INSTITUTO ESPAÑOL DE OCEANOGRAFIA (IEO)
UNIDAD DE INVESTIGACION DE CADIZ
MUELLE DE LEVANTE, S/N. PUERTO PESQUERO 11006 CADIZ
TEL:+34-56-261333   FAX: +34-56-263556
WWW: http://www.ieo.es

UNIVERSIDAD POLITECNICA DE CATALUÑA
VICERRECTORADO DE INVESTIGACION
AVDA. DR. GREGORIO MARAÑON, S/N 08028 BARCELONA
TEL:+34-3-4016111   FAX: 334-3-4016210
WWW: http://www.upc.es

UNIVERSIDAD DE BARCELONA
VICERRECTORADO DE INVESTIGACION
GRAN VIA DE LES CORTS CATALANES, 585 08007 BARCELONA
TEL: 34-3-3184266   FAX: 34-3-4125521
WWW: http://www.ub.es

INSTITUTO NAC. DE INV. Y TEC. AGRARIA Y ALIMENTARIA (INIA)
CENTRO DE INVESTIGACION Y TECNOLOGIA
AVDA. PADRE HUIDOBRO, KM. 7 28071 MADRID
TEL: +34-1-3476857   FAX: 34-1--3572293/3471472

INSTITUTO DE SALUD CARLOS III
FONDO DE INVESTIGACION SANITARIA
ANTONIO GRILO, 10 28015 MADRID
TEL: 34-1-5421800/5471177   FAX: 34-1-5421432
WWW: http://www.isciii.es

UNIVERSIDAD AUTONOMA DE MADRID
DPTO. FINANCIACION E INVESTIGACION COMERCIAL
DIRECCION: CANTOBLANCO; FAC. CC. ECONOMICAS 28049 MADRID
TEL: +34-1-3974300
WWW: http://www.uam.es

MINISTERIO DE AGRICULTURA, PESCA Y ALIMENTACION
COMISION COORDINADORA DE INVESTIGACION AGRARIA
JOSE ABASCAL, 56 28071 MADRID
TEL: +34-1-4423199

INSTITUTO DE FOMENTO DE ANDALUCIA
TORNEO, 26 41002 SEVILLA
TEL: 34-5-4900016

DIRECCION GENERAL DEL INSTITUTO
NACIONAL DE METEOROLOGIA
CAMINO MORERAS, S/N; CIUDAD UNIVERSITAR. 28040 MADRID
TEL: +34-1-5819630

DIRECCION GENERAL DEL INSTITUTO GEOGRAFICO NACIONAL
GENERAL IBAÑEZ DE IBERO, 3 28003 MADRID
TEL: 34-1-5979411/5333800
WWW: http://www.geo.ign.es

COMISION PERMANENTE DE LA CICYT
SECRETARIA GENERAL DEL PLAN NACIONAL DE I+D
ROSARIO PINO, 14-16 28071 MADRID
TEL: +34-1-3360400   FAX: +34-1-3360435
E-MAIL: sgpnid@cicyt.es

CENTRO PARA EL DESARROLLO TECNOLOGICO INDUSTRIAL
PASEO DE LA CASTELLANA, 141 28071 MADRID
TEL: +34-1-5815500   FAX: 34-1-5815576
E-MAIL: cmm@cdti.es
WWW: http://www.cdti.es

DIRECCION GENERAL DEL INSTITUTO GEOGRAFICO NACIONAL
SECRETARIA GENERAL
GENERAL IBAÑEZ IBERO, 3 28071 MADRID
TEL: +34-1-5333800

CENTRO ESTUDIOS Y EXPERIMENTACION
OBRAS PUBLICAS (CEDEX)
ALFONSO XII, 3 28014 MADRID
TEL: +34-1-3357500-5397251   FAX: +34-1-5280354
WWW: http://www.cedex.es

INSTITUTO TECNOLOGICO GEOMINERO DE ESPAÑA
CENTRO: INSTITUTO TECNOLOGICO GEOMINERO DE ESPAÑA
DIRECCION: RIOS ROSAS, 23 28003 MADRID
TEL: +34-1-3495700   FAX: +34-1-4426212

UNIVERSIDAD AUTONOMA DE BARCELONA
CAMPUS UNIVERSITARIO; EDIFICIO A 08193 BELLATERRA
TEL: 34-3-5811000/1101   FAX: 34-3-5812000
WWW: http://www.uab.es

UNIVERSIDAD POLITECNICA DE MADRID
AVDA. RAMIRO DE MAEZTU, 7 28040 MADRID
TEL: 34-1-3366052/3366051   FAX: 34-1-3366210
WWW: http://www.upm.es

FUNDACION COTEC PARA LA INNOVACION TECNOLOGICA
C/ MARQUES URQUIJO, 26
28008 MADRID
TEL: 34-1-5590881

ESIN - ESTUDIOS INSTITUCIONALES, S.L.
C/ ORENSE 68
28020 MADRID
TEL: 34-1-5721300   FAX: 34-1-570 08 09
E-MAIL: esin@tsai.es

PORTUGAL

INSTITUTO GEOLÓGICO E MINEIRO (IGM)
RUA ANTÓNIO ENES - 7, 1000 LISBON
TEL: 351-1-3529103/4/5 FAX: 351-1-525913   TELEX: 62195

INSTITUTO HIDROGRÁFICO (IH)
RUA DAS TRINAS - 49, 1200 LISBON
TEL:+351-1-3955119/24 FAX:+351-1-3960515   TELEX:65990 HIDROG P

INSTITUTO DE INVESTIGAÇÃO
CIENTÍFICA TROPICAL (IICT)
RUA DA JUNQUEIRA - 86, 1300 LISBON
TEL:+351-1-3645071 /3644946/7/3645518 FAX:+351-1-3631460

INSTITUTO DE METEOROLOGIA (IM)
RUA C DO AEROPORTO, 1700 LISBON
TEL:+351-1- 8483961   FAX:+351-1-802370   TELEX: 12352

INSTITUTO NACIONAL DE ENGENHARIA
E TECNOLOGIA INDUSTRIAL (INETI)
(CAMPUS DO LUMIAR)
ESTRADA DO PAÇO DO LUMIAR, 1699 LISBON CODEX
TEL:+351-1-716 5141/4211/51 81 FAX: +351-1716 09 01

INSTITUTO NACIONAL DE INVESTIGAÇÃO AGRÁRIA (INIA)
RUA DAS JANELAS VERDES - 92, 1200 LISBON
TEL.: +351-1-3951559   TELEX:+351-1-3977086

INSTITUTO NACIONAL DE SAÚDE "DR. RICARDO JORGE" (INSA)
AV. PADRE CRUZ, 1699 LISBON
TEL.:+351-17577070   FAX+351-1-7590441

INSTITUTO PORTUGUÊS DE INVESTIGAÇÃO MARÍTIMA (IPIMAR)
AV. BRASÍLIA, 1400 LISBON
TEL.:+351-1-3010814   FAX:+351-1-3015948   TELEX: 15857 IPIMAR P

INSTITUTO TECNOLOGICO E NUCLEAR (ITN)
ESTRADA NACIONAL Nº 10, 2685 SACAVÉM
TEL.:+351-1-9550021   FAX:+351-19550117;   TELEX:12727 NUCLAB P

AGÊNCIA DE INOVAÇÃO, S.A.
AV. DOS COMBATENTES, N° 43 - 10°/C-D
1600 LISBON
TEL: +351-1-727.16.21/77   FAX: +351-1-727.17.33

RUA DE SAGRES, 11
4150 PORTO
TEL: +351-2-610.33.59 / 60   FAX: +351-2-610.33.61

JNICT
AV. D. CARLOS I, 126, 1°, 1200 LISBON
TEL.: +351-1-397 90 21   FAX: +351-1-60 74 81
E MAIL: geral@jnict.pt

FEDERAÇÃO PORTUGUESA DAS ASSOCIAÇÕES
E SOCIEDADES CIENTÍFICAS
R. DA ESCOLA POLITÉCNICA, 58
1000 LISBON

MINISTRY FOR SCIENCE & TECHNOLOGY
PRAÇA DO COMÉRCIO, ALA ORIENTAL
1194 LISBON CODEX
TEL..:+351 - 1- 8812000   FAX:+351- 1 - 8882434

UNIVERSIDADE DE LISBOA
ALAM. DA UNIVERSIDADE,CIDADE UNIVERSITÁRIA
CAMPO GRANDE, 1699 LISBON CODEX
TEL.: +351-1-796 76 24   FAX: +351-1-793 36 24

SPI (SOCIEDADE PORTUGUESA DE INOVAÇAO)
EDIFICIOS "LES PALACES"
RUA JÚLIO DINIS, 242-2°-208 - 4050 OPORTO
TEL: +351-2-6099152   FAX:+351-6099164

GREECE

GENERAL SECRETARIAT FOR RESEARCH AND TECHNOLOGY
TEL: +30-1-6911122   FAX:+30-17711205
E-MAIL: postmaster@mhs-gw.gsrt.epmsh.gr

NATIONAL CENTER FOR SCIENTIFIC RESEARCH DEMOKRITOS
AGIA PARASKEVI, ATHENS
TEL: +30-1-6522965   FAX: +30-1-6522965

NATIONAL OBSERVATORY OF ATHENS
ATHENS
TEL:+30-1-3464161   FAX:+30-1-3421019

NATIONAL HELLENIC RESEARCH FOUNDATION (NHRF)
ATHENS.
TEL:+30-1-7217956   FAX:+30-1-7246618

FOUNDATION FOR RESEARCH AND TECHNOLOGY (FORTH)
HERAKLIO, CRETE, GREECE
TEL: +30-81-391500   FAX:+30-81-391555

INSTITUTE FOR LANGUAGE AND SPEECH PROCESSING (ILSP)
NEO PSYHIKO.
TEL:+30-1-6712250   FAX:+30-1-6471262

HELLENIC PASTEUR INSTITUTE
ATHENS
TEL:+30-1-6446143   FAX:+30-1-64223498

NATIONAL CENTER FOR SOCIAL RESEARCH (NCSR)
ATHENS
TEL:+30-1-3211477   FAX:+30-1-3636747

INSTITUTE OF MARINE BIOLOGY OF CRETE (IMBC)
HERAKLIO, CRETE, GREECE
TEL: +30-81-241992   FAX: +30-81-241882
NATIONAL CENTER FOR MARINE RESEARCH (NCMR)
ATHENS
TEL: +30-1-9811713   FAX:+30-1-9833095

INSTITUTE OF CHEMICAL ENGINEERING
& HIGH TEMPERATURE CHEMICAL PROCESSES
PATRA, GREECE
TEL:+ 30-61-990986   FAX:+30-61-990987

CHEMICAL PROCESSES ENGINEERING
RESEARCH INSTITUTE (CPERI)
THESSALONIKI, GREECE
TEL: +30-31-474128/9   FAX: +30-31-474121
GREEK ATOMIC ENERGY COMMISSION (GAEC)
AGHIA PARASKEVI ATTIKI
10559 ATHENS
TEL: +30-1-6515194   FAX:+30-1-6533939

NATIONAL DOCUMENTATION CENTER (NDC)
AVE. VASILEOS KONSTANTINOU 48. 11635 ATHENS
TEL: +30-1-7246825   FAX: +30-1-72446824

Technology Parks

ATTICA TECHNOLOGICAL PARK "LEYCIPPUS"
AG. PARASKEVI PO.BOX:60228
TEL: +30-1-6546637   FAX: +30-1-6536531

TECHNOLOGICAL PARK THESSALONIKI
THERMI THESSALONIKI PO.BOX:328
TEL: +30-31-471401   FAX: +30-31-471400

CRETE SCIENTIFIC AND TECHNOLOGICAL PARK
IRAKLION 711 10 PO.BOX:1447
TEL: +30-81-391900   FAX: +30-81-391906

PATRAS SCIENTIFIC PARK
STADIOU ST.
26500 PLATANI PATRA
TEL: +30-61-994046   FAX:+30-61-994106/661

ITALY

MINISTERIO DELL'UNIVERSITA E DELLA RICERCA
SCIENTIFICA E TECNOLOGICA (MURST)
PIAZZA JOHN F.KENNEDY, 20
I-00144 ROME
TEL: +39-6-59911

CONSIGLIO NAZIONALE DELLE RICERCHE (CNR)
PIAZZALE ALDO MORO, 7
I-00185 ROME
TEL: +39-6-6840031

ISTITUTO NACIONALE DI FISCIA NUCLEARE (INFN)
PIAZZA DEI CAPRETTARI, 70
I-00100 ROME
TEL: +39-6-6840031
CONSORZIO INTERUNIVERSITARIO
PER LA FISCIA DELLA MATERIA (INFM)
VIA DODECANESO, 33
I-16146 GENOVA
TEL: +39-10-3536357

ENTE PER LE NUOVE TECNOLOGIE,
L'ENERFIA E L'AMBIENTE (ENEA)
LUNGOTEVERE THAON DI REVEL, 76
I-00196 ROME
TEL: +39-6- 36271

ISTITUTO SUPERIORE DI SANITA (ISS)
VIALE REGINA ELENA, 299
I-00161 ROME
TEL: +39-6-49901

AGENZIA SPAIALE ITALIANA (ASI)
VIA DI VILLA PATRIZI, 13
I-00161 ROME
TEL: +39-6- 4404498

ISTITUTO NAZIONALE DI STATISTICA (ISTAT)
VIA CESARE BALBO, 16
I-00184 ROME
TEL: +39-6-3232617

AGENZIA PER LA PROMOZIONE DELLA RICERCA EUROPEA (APRE)
VIA FLAMINIA, 43
I-00196 ROME
TEL: 39-1-3232617

ASSOZIAZIONE ITALIANA PER
LA RICERCA INDUSTRIALE (AIRI)
VIALE GORIZIA, 25C
I-00198 ROME
TEL: +39-6-884883112

APPENDIX III
ACRONYMS*

ACL (P)	*Academia das Ciências de Lisboa* (Lisbon Academy of Sciences)
ANEP (S)	*Agencia Nacional de Evaluación y Prospectivas* (National Agency for Evaluation and Prospects)
ARTT (G)	Association for Research, Technology and Training
ASI (I)	*Agenzia Spaziale Italiana* (Italian Space Agency)
CDTI (S)	*Centro de Desarrollo Tecnológico Industrial* (Centre for Technological Industrial Development)
CECAM	European Atomic and Molecular Computing Centre
CEDEX (S)	*Centro de Estudios y Experimentación de Obras Públicas* (Centre for Studies and Experiments in Public Works)
CERN	European Organisation for Nuclear Research
CICYT (S)	*Comisión Interministerial de Ciencia y Tecnología* (Interministerial Commission on Science and Technology)
CIEMAT (S)	*Centro de Investigación Energética, Medioambiental y Tecnológica* (Centre for Energy, Environmental and Technological Research)
CIENCIA (P)	*Criação de Infraestruturas Nacionais de Ciência, Investigação Desenvolvimento* (Development of National Infrastructure for Science, Research and Development)
CIPE (I)	*Comitato Interministeriale per la Programmazione Economica* (Interministerial Committee for Economic Planning)
CIRIT (S)	*Comisión Interdepartamental de Investigación e Innovación Tecnológica* (Interdepartmental Commission for Technological Research and Innovation)
CNR	*Consiglio Nazionale delle Ricerche* (National Research Council)
COCEDE (P)	Committee responsible for the co-ordination of Portuguese participation in R&D programmes in the context of the European Union and the OECD
CPERI (G)	Institute of Chemical Engineering Processes
CSCT (P)	*Conselho Superior da Ciência e Tecnologia* (Higher Council for Science and Technology)
CSIC (S)	*Consejo Superior de Investigaciones Científicas* (Higher Council for Scientific Research)
CYTED (S)	*Programa Iberoamericano de Ciencia y Tecnología* (Latin American Science and Technology Programme)
DGICYT (S)	*Dirección General de Investigación Científica y Tecnológica* (General Directorate for Scientific and Technological Research)

* G=Greece; I=Italy; P=Portugal; S=Spain.

ECART	European Consortium for Agricultural Research in the Tropics
EKVAN (G)	Research Consortium for Improvement of Competitiveness
ENEA (I)	*Ente per le nuove tecnologie, l'energia e l'ambiente* (National Agency for New Technologies, Energy and Environment)
ENEL (I)	*Ente nazionale eneria elettrica* (National Electricity Agency)
EPET (G)	Operational Programme for Research and Technology
ERCIM	European Research Consortium in Computers and Applied Materials
ERD	Expenditure on R&D
ERDF	European Regional Development Fund
ESA	European Space Agency
ESF	European Structural Funds
EU	European Union
EUREKA	Programmes to Promote and Facilitate Industrial, Technological and Scientific Co-operation
FIS (S)	*Fondo de Investigaciones Sanitarias* (Fund for Health Research)
FORTH (G)	Foundation for Research and Technology
GDP	Gross Domestic Product
GSRT (G)	General Secretariat for Research and Technology
IAC (S)	*Instituto Astronómico de Canarias* (Astronomic Institute of the Canary Islands)
IACM (G)	Institute for Applied and Computer Mathematics
IBEROEKA (S)	Innovation Projects in the *CYTED* programme
ICE (G)	Institute of Chemical Engineering
ICS (G)	Institute of Computer Science
ICSI	International Computer Science Institute of Berkley
IEO (S)	*Instituto Español de Oceanografía* (Spanish Institute of Oceanography)
IESL (G)	Institute for Electronic Structure and Laser
IGM (G)	Geologic Mining Institute
IGN (S)	*Instituto Geográfico Nacional* (National Geographic Institute)
IICT (P)	*Instituto de Investigação Científica Tropical* (Institute for Scientific Tropical Research
IIMS (I)	*Istituto italiano di medicina sociale* (Institute of Social Medicine)
ILSP (G)	Language Processing Institute
IMBB (G)	Institute for Molecular Biology and Biotechnology

IMBC (G)	Marine Biological Institute of Crete
IMS (G)	Institute of Mediterranean Studies
INE (S)	*Instituto Nacional de Estadística* (National Institute for Statistics)
INETI (P)	National Institute of Engineering and Industrial Technology
INFN (I)	*Istituto nazionale di fisica nucleare* (National Institute of Nuclear Physics)
ING (I)	*Istituto nazionale di geofisica* (Geophysical National Institute)
INIA (I)	*Instituto Nacional de Investigación y Tecnología Agraria y Alimentaria* (National Institute for Agricultural and Food Research and Technology)
INIA (P)	*Instituto Nacional de Investigação Agrária* (National Agricultural Research Institute)
INM (S)	*Instituto Nacional de Meteorología* (National Institute for Meteorology)
INO (I)	*Istituto nazionale di ottica* (National Institute of Optics)
INSEAN (I)	*Istituto nazionale per studi de esperienze di architettura navale* (National Institute of Studies and Experiements in Naval Architecture)
INTA (S)	*Instituto Nacional de Técnica Aeroespacial* (National Institute for Aerospatial Technology)
INVOTAN (P)	Committee responsible for Portuguese participation in scientific research programmes promoted by NATO
ISCO (I)	*Istituto nazionale per lo studio della congiuntura* (National Institute for the Study of Economic Trends)
ISFOL (I)	*Istituto per lo sviluppo della formazione professionale dei lavoratori* (Institute for the Development of Vocational Training for Workers)
ISPE (I)	*Istituto de studi per la programmazione economica* (Institute for the Study of Economic Planning)
ISPESL (I)	*Istituto superiore per la prevenzione e la sicurezza del lavoro* (Higher Institute for the Prevention of Accidents and Safety at Work)
ISPT (I)	*Istituto superiore delle poste e telecomunicazioni* (Higher Institute of Posts and Telecommunications)
ISS (I)	*Istituto superiore di sanità* (Higher Institute of Health)
ISTAT (I)	*Istituto centrale di statistica* (Central Institute of Statistics)
ITGE (S)	*Instituto Tecnológico Geominero de España* (Technological Institute of Geomining of Spain)
ITN (P)	*Instituto Tecnológico e Nuclear* (Technological and Nuclear Institute)
JNICT (P)	*Junta Nacional de Investigação Científica e Tecnológica* (The National Council for Scientific and Technological Research)

LEST	European Solar Telescope
LNIV (P)	*Laboratório Nacional de Investigação Veterinária* (National VeterinaryResearch Laboratory)
MADESS	Materials and Devices for Solid State Electronics
MIDAS (S)	*Movilización de la Investigación, el Desarrollo y las Aplicaciones en Superconductores* (Mobilisation of Research, Development and Application of Superconductors)
MINER (S)	*Ministerio de Industria y Energía* (Ministry for Industry and Energy)
MOPT (S)	*Ministero de Obras Públicas y Transport* (Ministry for Public Works and Transport)
MURST (I)	*Ministero dell'università e della ricerca scientifica e tecnologica* (Ministry for the University and Scientific and Technological Research)
NASA	National Aeronautics and Space Administration
NCMR (G)	National Centre forMaritime Research
NHRF (G)	National Hellenic Research Foundation
ODP (I)	Ocean Drilling Programme
OECD	Organisation for Economic Cooperation and Development
OGS (I)	*Osservatorio geofisico sperimentale* (Experimental Geophysical Observatory)
OTRI (S)	*Oficina de Transferencia de Resultados de Investigación* (Office for the Transfer of Research Results)
OTT (S)	*Oficina de Transferencia Tecnológica* (Office for Technology Transfer)
PATI (S)	*Plan de Actuación Tecnológica Industrial* (Plan for Technological Industrial Action)
PAVE (G)	Programme for the Development of Industrial Research
PEDAP (P)	*Programa Específico de Desenvolvimento da Agricultura Portuguesa* (Specific Programme for the Development of Portuguese Agriculture)
PEDIP (P)	*Programa Específico de Desenvolvimento da Indústria Portuguesa* (Specific Programme for the Development of Portuguese Industry)
PETRI (S)	*Programa de Estímulo a la Transferencia de Resultados de Investigación* (Programme for the Stimulation of the Transfer of Research Results)
PITIE (P)	Programme of Integration of the Technology and the Electronics
PROBIDE (P)	Programme for the Development of Equipment
PRODEP (P)	*Programa de Desenvolvimento Educativo para Portugal* (Programme for the Educational Development in Portugal)
RAISA (I)	Advanced Research for Innovation in the Agricultural System

S&T	Science and Technology
SGPN (S)	*Secretaría General del Plan Nacional* (General Secretariat of the National Plan)
SINPEDIP (P)	System of *PEDIP* Initiatives
SME	Small and Medium Sized Enterprise
STRIDE	Science and Technology for Regional Innovation and Development in Europe
TBP	Technological Balance of Payments
VALUE	Valuation and Use of R&D in Europe